THE
WOODS HOLE
CANTATA

THE WOODS HOLE CANTATA

Essays on Science and Society

Gerald Weissmann

Foreword by Lewis Thomas

DODD, MEAD & COMPANY

New York

Published by Dodd, Mead & Company, Inc.
79 Madison Avenue, New York, N.Y. 10016

Distributed in Canada by
McClelland and Stewart Limited, Toronto

Manufactured in the United States of America

Designed by Claire Counihan

2   3   4   5   6   7   8   9   10

Library of Congress Cataloging in Publication Data

Weissmann, Gerald.
The Woods Hole Cantata.

Bibliography: p.
1. Biology—Addresses, essays, lectures.   2. Biology—
Social aspects—Addresses, essays, lectures.   3. Science
—Addresses, essays, lectures.   4. Science—Social aspects
—Addresses, essays, lectures.   I. Title.
QH311.W4   1985   574   85-4498
ISBN 0-396-08618-7

The author thanks the publishers of *Hospital Practice* magazine for permission to reprint the essays in this book, some of which appear here in somewhat altered form.

The essays "AIDS and Heat" and "The Urchins of Summer" also appeared in a much abbreviated form on the Op-Ed page of *The New York Times*.

Grateful acknowledgment is made to use the following material:

William Carlos Williams, *The Doctor Stories*. Copyright © 1932, 1933, 1934, 1937, 1938, 1941, 1943, 1947, 1948, 1950, 1951, 1962 by William Carlos Williams. Reprinted by permission of New Directions Publishing Corporation. William Carlos Williams, *Paterson*. Copyright © 1946, 1948, 1949, 1951, 1958 by William Carlos Williams. Copyright © 1962 by Florence Williams. Reprinted by permission of New Directions Publishing Corporation. William Carlos Williams, *Collected Earlier Poems*. Copyright 1938 by New Directions Publishing Corporation. Reprinted by permission of New Directions Publishing Corporation. William Carlos Williams, *Collected Later Poems*. Copyright 1944, 1948 by William Carlos Williams. Reprinted by permission of New Directions Publishing Corporation. William Carlos Williams, *The Autobiography of William Carlos Williams*. Copyright 1948, 1951 by William Carlos Williams. Reprinted by permission of New Directions Publishing Corporation. Erwin Panofsky, *Meaning in the Visual Arts*. Permission granted by Mrs. Gerda Panofsky. Excerpts from *Selected Poems*, Copyright 1947 by Bertolt Brecht and H.R. Hays; renewed 1975 by Stefan S. Brecht and H.R. Hays. Reprinted by permission of Harcourt Brace Jovanovich, Inc. Excerpts from *Man the Unknown* by Alexis Carrell, copyright 1935, 1939 by Harper & Row, Publishers Inc. Reprinted by permission of Harper & Row, Publishers Inc. Lionel Trilling, excerpt from *The Liberal Imagination*. Copyright 1950 by Lionel Trilling; copyright renewed in 1978 by Diana Trilling and James Lionel Trilling. Foreword copyright © 1976 by Diana Trilling and James Lionel Trilling. Reprinted with the permission of Charles Scribner's Sons. Excerpts from "A Toast," "Voltaire at Ferney," "New Year's Letter of 1939," "Ode to Terminus," "Bestiaries Are Out," and "After Reading a Child's Guide to Modern Physics," from W.H. Auden's *Collected Poems*, edited by Edward Mendelson. Permission granted by Random House, Inc. Excerpt from *The Horseman on the Roof*, by Jean Giono, translated by Jonathan Griffin. Permission granted by Random House, Inc. Excerpt from *Madness and Civilization: A History of Insanity in the Age of Reason*, by Michel Foucault, translated by Richard Howard. Permission granted by Random House, Inc. Excerpt from *Playing God: Genetic Engineering and the Manipulation of Life*, by June Goodfield. Permission granted by Random House, Inc. Excerpt from *In the Freud Archives*, by Janet Malcolm. Permission granted by Alfred A. Knopf, Inc. Arthur Koestler, *The Place of Value in a World of Facts: 14th Nobel Symposium*. Permission

To Ann,
My First Reader

Contents

CONTENTS

Foreword

In his professional, workaday life, Dr. Weissmann is as specialized a specialist as you are likely to find anywhere in the fragmented, compartmentalized world of modern medicine. He is a certified, card-carrying Rheumatologist, long trained and skilled in the diagnosis and treatment of arthritis and its connected disorders. When such a man settles down to write essays you might expect to be reading a lot about human joints, but not much else.

It is the commonest complaint about medicine these days that doctors in general have turned into doctors in particular, each knowing much more about much less. Indeed, it is believed that doctors in general have nearly vanished. Today's physicians, so it goes, are unable to look at their patients as whole human beings, seeing only the dimly outlined surrounds of a particular organ or tissue on which their attention is forever focussed. And as for the world at large, it does not exist.

Scientists are supposed to be even worse, more narrowly

concentrated at their benches, and Weissmann is himself a working researcher with a long bibliography of scientific papers trailing after him, presumably keeping him from thinking of much beyond the biochemistry and cellular biology of the inflammatory reaction. The reader's expectations might rise a bit at this news, but only as his heart sinks at the thought of still more about still less. Scientists are supposed to be the most superspecialized of reductionists, lost in the intricate details of their arcane puzzles, speaking languages beyond general comprehension, obsessed. Accordingly, a volume of reflections by someone with Weissmann's history might be expected to deal with technical matters of deep complexity, maybe good reading for other scientists in the same field.

It is not so. These essays range all over the place, touching human experience everywhere, from Poussin to autoimmune reactions to hospital architecture to William Carlos Williams to the lives of sea urchins to the annual Stockholm ceremonies to Vivaldi and back again, turning the two-culture cliché on its head.

I have known Gerald Weissmann since he came to Bellevue as a resident on the medical division for which I was responsible, about twenty-five years ago. I like to think that Bellevue is responsible, at least in part, for the length of his reach and the width of the net he casts. He has been a working part of this immense institutional model of the human universe for a long time, long enough for Bellevue to go to his head and stay there, as it does to the minds of most people who work in that great place for long stretches. What comes of it is an ungovernable curiosity about people of all sorts, a degree of modified hopefulness for the human condition, and a steadying affection for the species. Also, a driving wish to learn more, which means reading widely. Bellevue produces good readers, I have observed over the years. And, here, a good writer.

LEWIS THOMAS

Preface

T ODAY, while the viability of God and the novel are in question, the essay remains very much alive and well. Its Baconian tradition has proved particularly congenial to my colleagues in the sciences, whose far-from-amateur efforts have won them a readership that extends far beyond the limits of their discipline; I refer, of course, to Lewis Thomas, Stephen Jay Gould, Freeman Dyson, Harold Morowitz, Robert Coles, Peter Medawar, and to the man who started the flap about the two cultures, C. P. Snow. Snow is the caution here; although his novels can be read as warmed-over Trollope, the essays have changed the discourse of our time. Trained in measure and number, limited in poetic imagination by attention to detail, confined by logic and algorithm, the experimental scientist turns naturally to that most accessible of literary genres: the essay. Not many of us have ventured on the wilder shores of the lyric or the novel.

In point of fact, I have always wondered why poetry,

novels, and drama have been transformed by successive waves of revolutions with respect to form and function while the essay has remained pretty much unchanged. I have never heard anyone refer to an "avant-garde essay," in the sense that one talks about the formal adventures of Pound, Joyce, or Beckett. Just as well, too, one could argue. Both the essay and the methods of modern science seem to have the necessity of invention built in, so to speak. We have not been compelled to abandon their traditional structure in order to accommodate our new sensibility.

Guided, perhaps in retrospect, by these considerations, the essays collected in this volume will be traditional in form and unfold themselves along objective lines. They have been written over the last eight or so years as bimonthly pieces for the magazine *Hospital Practice,* whose editor, David W. Fischer, anticipated the recent turn to medical humanities by including a variety of nontechnical writings in his very successful venture. He found me at Bellevue Hospital where I spend most of the year doing medical research on inflammation and where I teach internal medicine to students of New York University. In the summers, at the Marine Biological Laboratory at Woods Hole, I work on marine creatures who, by virtue of simpler cellular arrangements, display in convenient form the origins of human pathology. In the course of this dual career, I have watched the transformation of medicine and biology from their base in a kind of arts-and-crafts movement that William Morris would have loved to their present development as the Jewels in the Crown of the modern scientific culture. This book represents, therefore, an effort to communicate with a larger audience some of the enthusiasm that this transformation has fired.

But I also present these essays as a personal response to that outrageous discrepancy of our age between our technical expertise and our social arrangements. Right now, on the ground floor of my hospital, there are any number of acutely ill patients in the Adult Emergency Room waiting their turn for a bed. Some have been stabbed, many are drugged, some are drunk, all have fevers; a harried staff is busy with their care. They are suffering from the diseases of social pathology. At the same time, here on the sixteenth floor, a young house officer is performing a sensitive radioimmunoassay on the plasma of a patient even sicker than those down below. Using techniques of refined precision unheard of six months ago, he will decide why the patient's lungs are choked with fluid. When cured of his disease, the patient will return to prison.

Not only our social flaws, but even the proud achievements of the biological revolution—the deciphering of the genetic code, the description of the fine structure of cells, and the unraveling of neurochemistry, to name a few—have left our community of scientists a touch dissatisfied. It is obvious to most of us that we have just begun to realize how much it is that we do not know. Having discovered that each level of scientific understanding reveals equally new displays of ignorance, we have become so sobered by the encounter that we refuse to guess what is or is not knowable. Whereas the *philosophes,* those sparkling essayists of the Enlightenment—Diderot, d'Alembert, Condillac—paid lip service to what they did not know, they retained that admirable confidence of the Age of Reason which persuaded them that it was *possible* to define the boundaries of the unknown. They were sure that these borders would soon be crossed by the armies of fact. We may have been to the moon, and made DNA in a dish, but in our age—armed to the teeth by fact—we have lost the belief that we can predict what we need to know. It is difficult to say whether it is good news that we are about to be the first generation to enter, consciously, the Age of Ignorance.

Faced with a mass of data from which the most open-minded of our contemporaries are forced to conclude "I really don't know!", we seem to have mellowed in our quest for absolutes. As Robert Darnton has recently reminded us in *The Great Cat Massacre and Other Episodes in French Cultural History,* the *philosophes* were obsessed with husbandry of the "tree of knowledge"; they devoted great energy to articulating the angles at which each branch of human knowledge or activity diverged from the primal trunk. Their device was a table of organization that rivaled the management ladder of IBM in complexity.

I doubt that it will ever be possible to draw up such a tree again. Indeed, one theme that unites the individual essays in this book is a kind of chronic impatience I have with hieratic intellectual systems or codes of conduct. I am persuaded that these inhibit us from confessing ignorance. Frequently the admission of ignorance is only a sign of rhetorical amiability, occasionally the symptom of plain forgetfulness, rarely a refusal to get involved in controversy. I hope the reader will be able to spot these uses of ignorance in my book, and hope also that the essays will be found innocent of that most trivial reason for an admission of ignorance: failure to distinguish that which is not known to anyone from that which is not known to the writer. Forgive me also if you fail to spot those subterranean structures so beloved by our colleagues of the Modern

Language Association. My hunch is that neither overt connections nor covert substructures fit the empirical temper of medical science in our time. Our century may have abandoned the possibility of a tree of knowledge and settled for the immediate bird in the bush.

We are not the worse off, I should have thought, for lacking the topiary map. The fallacy of misplaced ignorance implies that it is within our power to outline the contours of the unknown. But descriptions of earlier attempts to define the unknown teach us the general inutility of those efforts. When d'Alembert trimmed the tree of knowledge—to use Robert Darnton's phrase—he divided the sciences of nature into Mathematics and into "Particular Physics." Among the latter's branches, we find Zoology, and one of its branches, Medicine. This field of ours was again subdivided into Hygiene (prevention), Pathology, Semiotics (diagnosis and prognosis), and Therapeutics. In the eighteenth century, medicine was anterior to semiotics, and so it remains for me today.

In 1938, the philosopher Martin Heidegger addressed the Society of Aesthetics, Natural Philosophy and Medicine of the University of Freiburg im Breisgau on *his* attempt to trim the tree of contemporary knowledge. He titled his essay "The Establishing by Metaphysics of the Modern World Picture," neglecting the point that the real "World Picture" of Germany included the actions of Heidegger. As rector of his university he had presided over the enforcement of Nazi racial laws, as philosopher over the unconscious arrogance of the metaphysician. Flaws of moral judgment would seem to disfigure most easily those who value abstract ideas over people, systems over facts, form over function. Drawing upon its roots in human pain, experimental medicine at its best can guard us against the colder values, and it is from its viewpoint that these essays have been written.

THE
WOODS HOLE
CANTATA

The Woods Hole Cantata

Each August, when pink mallows fill wetlands by the bay, the Woods Hole Cantata Consort gives its annual performance. In recent years, chorus and orchestra have ventured the Bach "Magnificat," Haydn's "Creation," Handel's "Alexander's Feast," and tonight, the "Gloria" of Vivaldi. The concert is given in the fieldstone Church of the Messiah, which presides over a snug churchyard almost at the verge of Vineyard Sound. Inside, the church has the trim, no-nonsense bearing of its nautical setting: bright timber work, well-hewn pews, and high brass. On the festive nights of these concerts it becomes the intersection of at least two cultures.

The performers and their audience are drawn chiefly from the scientific summer community of the Marine Biological Laboratory, with help from year-round residents and the occasional semi-professional soloist. Led by Elizabeth Davis, wife of Harvard's eminent microbial physiologist, the performers run in age from adolescence to retirement; many of their surnames are familiar to readers of *The Journal of Biological Chemistry, The New England Journal of Medicine,* and *The Proceedings of the National Academy of Sciences.* Twenty minutes before the performance, the audience is already tightly packed into the pews; children and limber postdocs squat in the aisles. Latecomers will have to sit on the brick steps or the green lawn to hear the music through open doors. The audience—relatives, friends, coworkers, students—fills the hall with the tribal buzz and chatter that one hears at class reunions or graduations. The air is laced with *pizzicati* of nervous laughter that I recognize from my children's first recitals at music school or their undergraduate theatricals. As the performers file in, I look about at the community gathered here. One can identify embryologists whose winter habitats range from Hawaii to Naples, biochemists from Northwestern to Stony Brook, physiologists from Seattle to the Cambridges, physicians from Duarte to Lund. Tomorrow, in the laboratory down the hill, some of their number will trail the flow of ions through the squid's giant axon or impale the eye of a crab; others will watch granules explode in the egg of a sea urchin. Some, I imagine, may simply look up from a journal in the library and stare, in sheer puzzlement, at the tossing sea. Tonight, though, we are all here *en famille* to celebrate the ancient ritual of music in concert.

A hush before the opening bars; then, in a rush of sound, we are surrounded by the throb of an amateur chorus in full summer voice:

Gloria in excelsis Deo!

2

This splendid noise and its reverberations move us along for a while. I follow the text, agreeing in principle with the general sentiment:

Et in terra pax hominibus
bonae voluntatis

The music churns gloriously on. Frail in attention, my thoughts move from melody to text—and beyond. It occurs to me how unique an occasion this evening presents— what a series of clashing interests have been, literally, harmoniously resolved. The "Gloria" is Catholic liturgy, the church Episcopalian. The composer was a priest, the majority of performers are probably freethinkers. Vivaldi is Art, the audience lives Science. Under the roof of one church sit Moslem with Jew, Indian with Pakistani, Harvard with Yale: a truce is obtained, the conditions for which seem to have eluded a good bit of mankind. Nowhere have I seen it recorded—to turn to the clash of the Two Cultures— that late August evenings in the Hamptons, in Big Sur, or Woodstock are devoted to responsive readings from Darwin's *Origin of the Species* or Watson's *Molecular Biology of the Gene*. Nor, to return to clerical considerations, have I heard that religious enthusiasts have sat in rapt attention as, in joyous phalanx, they chant sections of Diderot's *Rameau's Nephew*, H. L. Mencken's essays, or Jacques Loeb's *The Mechanistic Conception of Life*.

The chorus modulates to

Domine Deus, Rex coelestis
Deus Pater Omnipotens

and my wayward thoughts turn, almost irretrievably, to Jacques Loeb, whose spirit remains so alive in this community. I had, indeed, almost tripped over him on my way to the concert; his simple gravestone—marked only with name and dates (1859–1924)—lies in the churchyard just a few feet from the rear entrance. What irony that the ashes of Loeb should come to rest by the Church of the Messiah! For Loeb, according to his biographer, R. L. Duffus, was champion of the antispiritual, a freethinker

> who believed that "living organisms are machines and that their reactions can only be explained according to the same principles which are used by the physicists"; who declared that nature was a blind muddler, with whom "disharmonies and faulty attempts are the rule, the harmonically developed system the rare exception"; who taught that consciousness and free will are illusions; who reduced the lord of creation to the status of a chemical solution bubbling in a tube; and who recognized nowhere in the universe a purpose or a God.

Yet there in the churchyard were his remains, cheek by jowl with the more God-fearing families of this small New England village. Loeb lies among the Swifts, the Stuarts, and the Fays, close to the founders of Woods Hole biology: F. R. Lillie, C. O. Whitman, and E. B. Wilson. A considerable distance separates his remains from those of more recent additions to this necropolis of the eminent: Selman Waksman, Hans Einstein, and Stephen Kuffler. We owe the presence of these distinctly non-Yankee names to an enlightened Church and community, who have turned the little graveyard into an ecumenical cemetery of biology à la Père Lachaise. But Jacques Loeb in a churchyard? The supreme mechanist, Loeb was persuaded that

life, i.e., the sum of all life phenomena, can be un-
equivocally explained in physico-chemical terms.

In his dazzling career as an experimental biologist at
Strasbourg and Würzburg, then Bryn Mawr, Chicago,
Berkeley, and the Rockefeller Institute, spelled by long
summers at Woods Hole, he combined imaginative science
with a tough, antimetaphysical stance in matters of philosophy. He took as first principle that

> . . . our existence is based on the play of blind forces,
> and [is] only a matter of chance . . . We ourselves are
> only chemical mechanisms . . .

Loeb presented this view (printed as the first chapter of
The Mechanistic Conception of Life) at the high-water mark of
pre—World War I scientific optimism, the First International Congress of Monists at Hamburg in September of
1911. The meeting, a veritable "concourse of the omni-
scients," according to Donald Fleming, attracted "to this,
the most radical, libertarian, and anti-Prussian of German
cities, delegations of freethinkers, freemasons, ethical cul-
turists, socialists, pacifists, and internationalists . . ." And
to this audience of several thousand, Jacques Loeb outlined
the truths of the new quantitative biology, of which he was
so much a part. The work of Mendel and Morgan had established the mathematical rules of heredity, Perrin and
Millikan had experimentally shown the *existence* of atoms
and molecules, and Loeb himself had shown not only that
the instincts of lower animals could be redefined as "tropisms," but also that one could approach the challenge of
creating life in the lab by parthenogenesis. What need for
a sentient *Pater omnipotens,* when, as Loeb predicted in his
address, future scientists will doubtlessly achieve "the task

of producing mutations by physico-chemical means . . . and prove that they have succeeded in producing nuclear material which acts as a ferment for its own synthesis and thus reproduces itself." *

With biology firmly established on the quantitative terms of physics and chemistry, with behavior a predictable response to chemical tropisms, with the beginnings of artificial life sputtering at the bottom of a beaker, all that was left to explain was the place of values in this world of fact. Loeb asked:

> How can there be an ethics for us? The answer is, that our instincts are the root of our ethics and that the instincts are just as hereditary as is the form of our body. We eat, drink and reproduce not because mankind has reached an agreement that this is desirable, but because, machine-like, we are compelled to do so. . . . We struggle for justice and truth since we are instinctively compelled to see our fellow beings happy.

Loeb was convinced that both individual and group behavior followed from genetically programmed "instincts" as surely as the markings on the wings of fruit flies followed from the rules of Mendelian genetics. And genetics would soon have a chemical basis. Evil was simply a mutation or an uncommon error:

> Economic, social and political conditions or ignorance and superstition may warp and inhibit the inherited instincts

---

* Loeb's prophecies were fulfilled when Hermann J. Müller induced mutations by X-rays in 1926; the blueprints of the nuclear material and its reproduction were drafted by Francis Crick and James Watson in 1953, and the ferment (DNA polymerase) was described by Arthur Kornberg in 1955. These discoveries were rewarded in the most appropriate manner, at Stockholm.

and thus create a civilization with a low development of ethics. Individual mutants may arise in which one or other desirable instinct is lost, just as individual mutants without pigment may arise in animals; and the offsprings of such mutants may, if numerous enough, lower the ethical status of a community . . . Not only is the mechanistic conception of life compatible with ethics: it seems the only conception of life which can lead to an understanding of the source of ethics.

It is difficult not to be charmed by this generous summary. Loeb here suggests not only that ethics derive from genetics but also that an inheritable instinct to make our fellow men happy guarantees the "struggle for truth and justice." Believing that a genetic blueprint specifies the virtue of our biological machine, Loeb is persuaded that "low ethical status" results from rare mutations or unusual "economic, social and political forces." This Panglossian view of the gene machine is, of course, as dated as the scientific optimism of the beginning of this century. The brutal wars and senseless murders which followed so closely upon the hopeful assembly at Hamburg have given sufficient evidence, I should have thought, that the direct opposite of Loeb's hypothesis is equally likely to be true. Verdun and Coventry, Dachau and Nagasaki make it just as likely that the blueprints of "instinct" specify brutish conduct and that the ethical drive for "justice and truth," which seemed so natural to Loeb, may constitute the aberrant mutation.

Loeb saw himself as heir to a rich patrimony of rational thought dating from the Enlightenment. Son of Benedict Loeb, a well-off Jewish importer who had settled in the Rhineland, he was raised in a secular, cultivated home of which the major intellectual heroes were the humanists of eighteenth-century France. Throughout his life, Jacques Loeb looked to the *philosophes* as his guide; indeed, his *The Organism as a Whole* (1916) was dedicated to Denis Diderot. And when the whirlwind struck, when the ship of scientific optimism was sunk by the guns of August, he remained true to the spirit of

that group of freethinkers, including Alembert, Diderot, Holbach and Voltaire, who first dared to follow the consequences of a mechanistic science, incomplete as it then was, to the rules of human conduct, and who thereby laid the foundations of that spirit of tolerance, justice and gentleness which was the hope of our civilization until it was buried under the wave of homicidal emotion which swept through the world in 1914.

I look about in the church as the chorus turns to the *Miserere*. A list of young Americans killed in that wave of homicidal emotion hangs on the wall. The music rises:

> *Qui tollis peccata mundi*
> *miserere nobis*

Loeb's public responses to the war constitute some of the more lucid formulations of liberal opposition to slaughter on the Western front. He said that the war made him "sick," and framed an emotional plea against the jingoism of the time in "Biology and War," which was published in *Science* on January 26, 1917.

The biology of which the war enthusiasts make use is essentially antiquated, and so we need not be surprised to find that they consider war to be based on what they call the "biological law of nature," or the "survival of the fittest."

8

The war enthusiasts also derive from what they are pleased to call the "law of nature" the statement that "superior races" have the right of impressing their civilization upon "inferior races." The information concerning the relative values of races is furnished by a group of writers who call themselves "racial biologists." This "racial biology" is based on quotations from the erudite statements of theologians, philologists, historians, politicians, anthropologists, and also occasionally of biologists, especially of the nonexperimental type . . . the sad fact remains that this pseudobiology has had at least a share in the production of the tragedy which is being enacted in Europe. For wars are impossible unless the masses are aroused to a state of emotionalism and fanaticism, and the pseudobiology of littérateurs and politicians *may serve this purpose in the future* as it has in the past [my italics].

Contrast this view, if you will, with the majority sentiment, as articulated by William Osler (See "Against *Aequanimitas*").

Loeb was provoked not only by the senseless war, but also by the racial nonsense chucked about by "pseudobiologists" on both sides. He had left Germany in 1891, partly because of his political and social opposition to Prussian orthodoxy and partly because the *numerus clausus* made it even more difficult than in America for a Jew to scale the academic ladder. He may, therefore, have been more sensitive than many to the waves of national intolerance that rolled over his adopted country as it entered, fought, and finally won the war.

The prodigious pace of Loeb's research did not slacken in wartime. Indeed, his work—which had won him international fame and gained him the general ear—became more exacting and more quantitative. But, in addition, during the war years of 1914–1918 he took up his pen to plead before a general audience the liberal ideals of pacifism, tolerance, and understanding, and to attack the misuse of poorly perceived biology. In 1914 came "Freedom of Will and War" in the popular *New Review;* "Science and Race" in the transient *Crisis;* in 1915 came "Mechanistic Science and Metaphysical Romance" in the *Yale Review;* in 1916, *The Organism as a Whole: From a Physico-Chemical Viewpoint;* in 1917 he published "Biology and War."

After 1917, however, there was only public silence from Loeb. Had he simply stopped trying? Were reason and scientific optimism insufficient weapons with which to fight the accelerating mischief? Why had the new Enlightenment not bettered the "economic, social and political conditions, or ignorance and superstition" which had dragged the most advanced countries of the West into self-destruction? One can sense from a letter to his friend, the great Swedish chemist Arrhenius, some of the despair Loeb felt in the xenophobic climate of postwar America:

> Politically I think America is in a bad way. We are suffering from a wave of reaction and from religious and racial fanaticism just as the European countries do . . . I wish something might be done to make the world a little more promising for the next years, but the outlook is bad. Personally I work hard because *I want to forget* and I am very grateful to science that it permits us to forget a good deal of the outside world [my italics].

It seems likely from this, and other letters, that the lessons of war changed Loeb in a profound way; and I think that I can document the consequences of that change by a little bookkeeping from his bibliography. The last year before the war, 1913, was a productive one for Loeb; it was the year that he and Reinhard Beutner correctly deduced that the potential electrical difference which exists across biological membranes might be maintained by phospholipid layers. In 1913, he published sixteen papers, and of these, fourteen described experiments with living animals. In 1920, an equally productive year, Loeb published eighteen papers but only two described research on living creatures. "I want to forget," wrote Loeb, and left his prickly sea urchins to launch studies on purified proteins *in vitro*. These important essays into colloidal chemistry were to occupy him until his death.

This drive to reductionism, which has characterized the biological revolution of our time, fits (ironically) Loeb's description of a "tropism." Tropisms have long been discredited as adequate explanations for the wide range of human behavior; they cannot, for example, explain our deepest psychic lives, our art, or our music. But to me, they seem sufficient to describe the more limited behavior of scientists at their job.

Loeb first encountered tropisms in 1888; their analysis established his reputation. Certain caterpillars emerge in the spring and move to the tips of branches in order to feed on the forming buds. This behavior, before Loeb, had been generally attributed to the playing out of an innate, almost metaphysical, drive for self-preservation. Loeb showed that this "instinct" had a distinct biological basis in photosensitivity. Since the perception of light is, by definition, a chemical event—a photoelectric cell can tell if a light is on or off—instincts are nothing but chemical reactions. Loeb found that if the only source of light with which he provided the caterpillars was from a direction opposite that of the food, the caterpillars would move to the direction of the light—and starve. Loeb interpreted these data on heliotropism to show that free will could be ruled out as motive for *any* action; all behavior had roots that could be reduced to biochemical responses. I view Loeb's experiment as the perfect description of what scientists do in a laboratory; they crawl in heliotropic wriggles to the source of light. Often that light shines from the wrong direction, but follow it they do, and with results anticipated by caterpillars.

Quoniam tu solus sanctus
tu solus Dominus

The Vivaldi is soaring to resolution. The church, warmer than at the beginning of the evening by virtue of summer, of an enthusiastic audience, and of modest ventilation, is filled with ravishing chords of the Baroque. I become persuaded by the thought that Loeb was wise, indeed, when he decided to forget a good deal of the outside world and to devote the last six or so years of his life to establishing the precise isoelectric point of gelatin. Admired by the freethinkers of his time (Thorstein Veblen, Sinclair Lewis, H. L. Mencken), Loeb and his generation of mechanists survive only in the memories of their students. Unfortunately Paul deKruif was one of these, and deKruif's popular accounts of Loeb made the mechanistic conception of life sound like a sophomoric attempt at village atheism. Mechanism did not outlast Prohibition.

But it is possible to argue that mechanistic philosophy, which drove Loeb and his generation into transient fallacies of social thought, has produced some of the noblest and

most lasting products of our civilization. By ac
the descriptions of man and nature can be red
blueprint of machines, our generation has bro
netic code and landed on the moon. As Loeb
predicted, by deciding that not only our acti
our thoughts, are governed by rules (struc
control mechanical or electronic machinery, v
formed the continents into matrixes of micro
social sciences into schools of structural eng
would have been surprised that this "tropis
and chemical models of the natural world has
to advance the sum total of human happin
1913. We now know, or at least believe, t
rational, anticlerical positions of Diderot
sophes, or of Loeb and the mechanists, are no more likely
to resolve the problems of war and violence than is the text
of the "Gloria." And while mechanistic philosophy may
describe adequately how science works, it does not offer
consolation for the world it produces: Vivaldi may be more
appropriate to that task. Nevertheless, whether we like it
or not, all experimental scientists are mechanists now.
Ravished by Art in this church tonight, we will wake in
the morning to work in the mills of Fact, the construction
of which we owe to the mechanistic conception of life.

Si monumentum quaeris, circumspice ("If you seek his monument, look about you") reads a tablet at the entrance to
the main building of the Marine Biological Laboratory. The
quotation, directed at the work of Christopher Wren, comes
from St. Paul's Cathedral. The tablet is dedicated to F. R.
Lillie, who lies near Loeb in the churchyard. The building
is named "Lillie," and it is at right angles to one called
"Loeb," where dozens of students are enrolled in the mod-

ern version of the physiology course that Loeb taught in the 1890s. Loeb's own tablet, near Lillie's, reads:

JACQUES LOEB
1859–1924

BRAIN PHYSIOLOGY
TROPISMS, REGENERATION
ANTAGONISTIC SALT ACTIONS
ARTIFICIAL PARTHENOGENESIS
COLLOIDAL BEHAVIOR

I am pleased that this community of science acts each day as if Diderot and Loeb were right, that we look at our jobs as the solution of one mechanical problem after another. I am also reassured that we are all here listening to Vivaldi, whose final notes cannot, by any stretch of the imagination, be ascribed to tropisms. Perhaps that is why the ashes of Loeb lie buried outside this monument to the nonmechanistic, the irreducible: this church filled with music and the students of Loeb's students. After years of teaching, of writing, of reasoning, the shade of Jacques Loeb may have earned the right to listen in quiet satisfaction to this harmony.

But tomorrow, when oscilloscopes flash in the lab, when the electrodes twitch with the signals of a nerve, when life precipitates from white threads of DNA, Loeb's monument will rise about us. The flesh may perhaps be weak, but the spirit of mechanistic science survives as long as one postdoc dips a sea urchin egg into a beaker of salt. What rest by the Church of the Messiah are the ashes of Loeb, the disappointed optimist. His monument is the mechanistic conception of life, the manifesto of a biological revolution which has spread far from the cozy waters of Woods Hole.

Cum sancto spiritu
in gloria Dei Patris.
Amen.

Bellevue: Form Follows Function

THESE days, most of the great teaching hospitals of the world resemble urban sites in search of an archaeologist. Perhaps their pavilions, wings, annexes, and towers were first planned in conformance with one central design. But thanks to a donor here, a social need there, an architect's vanity, or a committee's decree, these institutions astonish the modern visitor with a display of architectural styles as varied in structure as in function. Like the Necker and Saint Antoine hospitals in Paris, Guy's and the former Addenbrook's of England, the Hopkins and the Brighams of our own country, Bellevue Hospital of New York sends structural signals of history and confusion, change and accommodation.

The main building, which was conceived in the sixties and finally occupied in 1975 (Katz, Waisman, Weber, Strauss, Blumenkranz; and Bernhard, and Pomerantz & Breines), is a square block of pebbled concrete panels and glass twenty-five stories high. This fortresslike structure by the East River is a municipal version of its more graceful precursors in the Corporate American Empire style of lower Park Avenue. Indeed, its glass-encased reception lobby recalls the spacious dining room of the Four Seasons restaurant in Mies van der Rohe's Seagram building. With what appropriate irony! The homeless alcoholics of Bellevue frequently sleep off Seagram's Four Roses on the vinyl banquettes that stud this toned-down version of Mies' monument to the three-martini lunch.

No solitary megalith, our citadel of healing is connected by a series of arbitrary tunnels, walkways, and corridors to vestiges of earlier orders of architecture. One or two of the red brick and green copper pavilions of McKim, Mead and White's old Bellevue (1908–1916) still remain amongst the rubble of their cohort. Their abandoned loggias and decorated balconies recall the epoch when fresh air was still available and considered therapeutic. A large administration building (1939, also McKim, Mead and White) courageously blends elements of a Palladian facade with the muscular machine forms of the Depression. On top of this edifice are corrugated metal superstructures that shimmer in the winter sun. These boxlike affairs were used as temporary operating rooms in the fifties. *O tempora! O moiré!* In the days when house officers were truly "interns" and had to reside in hospital, this building housed their sleeping quarters and scruffy lounges where poker games began at midnight. In the basement was the doctors' dining room, where municipal waitresses served white dinners (filet of sole, cauliflower, macaroni) on white plates over white tablecloths to men in white. (There were very few women.)

The old C and D Building of the chest service, center for many years of tuberculosis treatment and research, also exposes its residual balconies to the Manhattan soot. It is of a piece and period (1939, York and Sawyer) with the administration tower. Its former use now superseded by streptomycin and isoniazid, the chest building houses a large security department, some squalid research warrens, and the prison ward. In corridors where once the phthisic wandered, with collapsed lungs, cell biologists and immunologists now compete for *Lebensraum* with manacled transfers from Riker's Island and bulky correction officers. The facade is an acceptable mix of neo-Georgian details with (what foresight!) the bulky elevations of Alcatraz.

Another survivor is the psychiatric building (1937, Thompson, Holmes, and Converse). This red brick and granite structure appears to have been based upon a misapprehension of the Villa Medici. Its windows, barred to foil the suicidal, contrast with *cinquecento* porticoes, Michelangelesque stairways, carved pediments, and fluted cornices—visual tributes to the rational use of the higher faculties. Psycho has played host to the inspired and the dangerous: from Norman Mailer to the East Side Slasher, from Malcolm Lowry to FLN bombers.

These major buildings house the 1,200 beds that have made Bellevue (founded as an almshouse in 1736) not only the oldest city hospital in our country, but also its largest teaching institution. And perhaps the first fact taught to student or visitor is that there is no logical way to enter it. Unless brought directly by ambulance to the emergency entrance (cleverly hidden behind a corral of parking lots), one has to follow a wacky trail indeed to the new Bellevue. From First Avenue, the main public thoroughfare that brings the visitor by taxi, bus, or foot (the nearest subway is four long blocks away), one is directed to a narrow walkway, covered by the sort of plastic canopy that is used to shield tobacco in the Connecticut River Valley. En route, the visitor passes a seven-tiered parking garage of scruffy concrete, which has clearly metastasized from a primary site in the Trenton of 1960. In keeping with the modern usage of teaching hospitals, the walkway leads to a side entrance of the old administration building. The front portico is bricked up.

Upon entering the makeshift lobby—an antechamber constructed from the confluence of older corridors—one is confronted by three paths: One can move into the coffee shop, a concession which has been designed in the neo-Formica style of 1970 Milan with later echoes of *l'ecole* Burger King. Or, one can head toward the old chest building. The corridor of this pathway was modernized in the late sixties by an admirer of Charles Moore, the exuberant architect of the New Orleans World Fair, whose major contribution to the postmodern style was the substitution of random arches or witty diagonals for the traditional square or rectangle. In Bellevue, this esthetic was achieved by snaking sausage-like strips of fluorescent light about scattered baffles dropped from a cocoa-colored ceiling. The remainder of this color scheme is beige and spinach green.

But let us follow the third, or main, route to the new hospital. Its architectural forms derive from the egalitarian devotions of Robert Venturi, who so admired the nocturnal cityscape of Las Vegas. The floor is dominated by five bright strips of colored carpet that intersect the gray length of the corridor. They are color-coded for the several stations of the patients' cross: blue for the outpatients, green for the children's playroom, orange for the screening nurse, and so on. At each station, a similarly colored banner proclaims the destination: Outpatients, Play Area, etc. Hung from the ceiling, the bright fabrics evoke the *souks* of Marrakesh.

At last one enters the new Bellevue and finds oneself in the Four Seasons. In this lofty lobby, one is finally in the presence of the Health and Hospitals Corporation! Banks of automated express and local elevators, efficient guards, and computer-assisted receptionists confront the visitor amidst blue carpets and polished terrazzo. One notes with dismay the underutilized main entrance designed by the architects of this great edifice. Its wide glass doors open rarely, since the entrance is inaccessible to any of the city's streets. An occasional respiratory therapist enters from a hidden path leading to the nursing school.

The ground floor promenade, however, is exhilarating for reasons other than its mélange of architectural eccentricity. For the scene is dominated by the crowds that move along its passages. We are surrounded by the life of our city. The classic Bellevue derelict, with gaping tennis shoes and fading pinstripe trousers, is flanked by a tidy covey of gays with dyed hair and chrome-studded black leather pants. A short, pregnant Hispanic woman is jostled by a crew of hefty suburban ambulance drivers sporting ROCKVILLE CENTRE quilted on the backs of their nylon jackets. A tiny Chinese scholar is accompanied by a West Side social worker in space shoes. Two youths in running clothes and stereo headsets accompany a third, already on crutches. A drug salesman in polyester bumps alongside a gaggle of bag ladies. House officers, in khaki and topsiders, mingle with shaven-headed Hare Krishnas. The mass is monitored by ubiquitous hospital police; the gun-toting officers toss pleasantries at a trim nurse or two.

And the signs! In English, Spanish, and Chinese—their messages, first embalmed in the plastic *sans serif* iconography that Massimo Vignelli designed for Bloomingdale's, have become modified by notices of clinic openings and closings inked out in Magic Markers. Long "Bills of Patients' Rights," declarations of "Pilferage Control," posters announcing staff lessons and rap sessions devoted to "Spanish," "Good Nutrition," "The Appreciation of Modern Art," and "Rape Victim Support" mingle with displays of athletic trophies won by the bowlers, runners, and softballers among the employees. Pass by all these, and, by means of the automatic elevators, one arrives at last at the real Bellevue—its wards and clinics.

On the upper floors, the aspect changes: Confusion is transformed into clinical order. The corridors are clean, wide, and glossy. Neat signs direct the visitor to spacious rooms that yield stunning views of the river or the Manhattan skyline. Activity at the nursing stations is brisk and efficient, monitors and intercoms scan the life of the wards. We can, if so inclined, experience an epiphany of enlightened social service: The indigent of New York—its domestics and derelicts, aged and addicted—are bedded in clean linen beneath private television receivers. Airy dayrooms occupy each corner of every floor, modern prints by Warhol and Indiana brighten clinics and examining rooms. The city has provided a spacious cafeteria and well-stocked libraries for both staff and patients. Rehabilitation has its pool, the intensive care units (coronary and general) are filled with gadgets appropriate to the maintenance of pulse and breath, radiology suites glisten with the diagnostic triumphs of the decade. In the streets below, pavements may crumble, the subways may rush to a halt, the garbage may spill into plugged gutters. Here at Bellevue, however, at least one city service has survived the recent federal war, not against poverty, but against the poor.

Bellevue functions as a kind of realization of the social optimism of the early sixties, when energy was cheap and the American Empire was infused with the vigor of Camelot. In those days before Vietnam, under Wagner and Lindsay, New York City had caused to be designed for its poor a structure in which to provide medical care that was to be no less splendid than those glass pyramids that housed its great corporations. I remember when the drilling of the new foundations began. At the time, the Department of Medicine was quartered in the old A and B Building, then only a shabby relic of the splendid design of McKim, Mead and White (1908). The open balconies on the river had long been abandoned or turned into tatty conference rooms. The fine marble stairwells had cracked. The manually operated elevators were frequently out of action; when not, the call buttons had the classic Bellevue notice affixed: "Ring Up for Down." At night, the operators being absent, students and interns alternated in running these rusty cages in order to move patients from the wards to X-ray or to other destinations.

The wards, owing to their 1908 design, were large and commodious, but during the winter months anywhere from thirty to forty beds were tiered in three parallel rows, sometimes spilling into the corridors. The windows were uniformly grimy. In order to permit easier access and observation, the beds of the sickest patients were moved toward the end with the nurses' station. Communication was poor and house officers sometimes napped next to a very ill patient in the early hours of the morning.

Since the B Wing was to the north, where the huge pillars of the new Bellevue were to be sunk, while the A Wing was remote from the excavation, cardiac auscultation during working hours was possible only in the A Wing. This state of affairs led Lewis Thomas (then Chairman of Medicine at NYU-Bellevue) to quip, "A decade of NYU students will have learned only one aspect of cardiology: B2 is louder than A2."

That phrase of yesteryear summons up not only the transient nature of our craft's terminology, but also the changes that time has worked with respect to disease. The old Bellevue of McKim, Mead and White was filled with patients who suffered from diseases that we no longer see in our new high rise by the river. In the three decades of my Bellevue career, the transitions of diagnosis have been as abrupt as those of the landscape. Then, as now, the major social disease was acute and chronic alcoholism with its devastating jabs at heart, liver, brain, and blood. But, during my student and house officer years, the beds of A and B were also filled with rheumatic fever. Valvular heart disease and endocarditis were found mainly among the rheumatics or the already declining population of patients with tertiary syphilis. Deforming gouty arthritis and tophaceous gout brought in quite a few, and each winter flooded the wards with pneumococcal pneumonias. Iron lungs—the awkward respiratory aids of our era—were stationed outside the pediatric isolation wards to aid in the ravages of polio. Tuberculosis accounted for the population of an entire building. Indeed, the old chestnut of Bellevue morning rounds was: "This patient is clearly suffering from cirrhosis of the liver, pulmonary tuberculosis, and tertiary syphilis. But what brought him into the hospital?"

Only in retrospect do we recognize that our patients, arranged so neatly in the tiers organized by the planners of 1908, were burdened by diseases that were rapidly waning in incidence. Thanks, in part, to experience at Bellevue and its sister institutions, rheumatic fever and syphilis have yielded to penicillin, tophaceous gout to probenicid and allopurinol, poliomyelitis to the vaccines of Salk and Sabin, and tuberculosis to the successors of streptomycin. But if these diseases accounted for so many admissions to the old hospital, why are the beds of new Bellevue still so full?

Well, to begin with, cancer and the degenerative diseases of heart, lung, and brain are still with us; their therapy (if not, alas, their cure) occupies the attention of man and machine. But it is also true that, like alcoholism, the infective insults of streptococci and tuberculosis are abetted by poverty, dirt, and urban crowding—what the young leftists of 1968 called the etiology of "diseases of oppression." And since there is little evidence that these social imperfections have been ameliorated, newer diseases of the underclass have begun to replace the old. The "poor" of McKim, Mead and White's era were the Irish, Italian, and Jewish immigrants of lower Manhattan. Indeed, it was clinical dogma—drawn from experiences at Bellevue and Boston city hospitals—that rheumatic fever was an "Irish disease"; this observation was not based on data adjusted for class or income. In today's Bellevue, the "poor" are black, Hispanic, Haitian, and Chinese. They also include a number of children of the white middle class who have migrated to Manhattan in quest of drugs or the sexual company of their own gender.

We now find our wards filled with cases of AIDS, "brown heroin kidney," "gay bowel," oral chancres, gonococcal arthritis, the hepatitis of drug abuse or pederasty, the endocarditis relayed by infected needles, and the acute respiratory distress that follows a jet of flaming cocaine. The ill effects of drug abuse and the consequences of modern sexual practices are beginning to edge out the effects of our traditional abuse of alcohol. We have so much to look forward to!

Here, then, a marriage between form and function in the new Bellevue! The punk-rock entrance to Bellevue welcomes a Rastafarian with infected needle tracks. The neon strips of the Venturi school light up the jaundiced eyes of a slim hairdresser. The psychedelic banner of the screening nurse greets the frightened teenager aflame with speed. And on the quiet corporate floors of the new hospital tower, the professional middle class (students, house officers, attending physicians) sorts out the ranks of those wounded in our social wars. The mugged, maimed, drugged, and infected are victims of epidemics as virulent as those that filled the gracious loggias of 1908. The high-rise columns and modular panels of corporate architecture now house the losers of new urban scrimmages, for whom there is no place in the similar towers of Wall Street or midtown. In both the old Bellevue and the new—and for this reason there is perhaps no more engaging hospital in which to work—we have been able to appreciate not only *what* happens in disease, but also to *whom*. The setting of our encounter was, and is, entirely appropriate. The forms of our buildings spell out the social litany of their function. It is often a sad and always a confusing refrain.

Foucault and the Bag Lady

On cold Mondays in February the victims of winter cluster at Bellevue. Infarction and pneumonia, exposure and gangrene drive the inhabitants of street and park to its doors. It was therefore a routine event when police brought the woman we'll call Mrs. Kahaner to the emergency room. Suffering from apparent frostbite, she had been found at the side entrance of Lord & Taylor early on Sunday morning and was admitted with a temperature of 93 degrees Fahrenheit.

The nurse's notes described her belongings. She was evidently the prototypical bag lady. Her rags, pots, and jumbled impedimenta overflowed five shopping bags crammed into a wire cart. An unkept appointment slip found in her purse identified her as a former patient in one of the New York State mental hospitals from which she had been last discharged three years ago. A telephone call to that hospital yielded her provisional diagnosis (chronic schizophrenia), her medications on discharge (the usual phenothiazine derivatives), and her age (fifty-six). The nurse had guessed Mrs. Kahaner to be an unkempt forty-five or so. The telltale areas of black gangrene on her fingertips had almost directed her admission to the surgical service, but a more thorough examination by an astute medical resident brought her to the medical wards, where our Rheumatology Service was called to see her the next morning.

The resident had observed that the skin over Mrs. Kahaner's face was drawn so tightly that she seemed to be peering out from behind a mask of fine leather. Over her chest and extremities, patches of pearl white skin alternated with areas of café-au-lait pigmentation. Her digital gangrene was only an extreme sign of impaired circulation—even after she had been warmed to normal temperature, her hands and feet remained mottled purple and pink. Her hands were also deformed by pincerlike contractures, her mouth was dry, and her eyelids crusted. X-rays showed fibrosis of the lungs, an enlarged right side of the heart with a prominent pulmonary artery segment, and diffuse calcification of the soft tissues. The admitting diagnosis was scleroderma.

As we examined her, Mrs. Kahaner remained resolutely silent. Warmed, hydrated, with oxygen aboard, she slowly responded to those requests necessary for physical diagnosis, but not at all to questions of medical or personal history. Frightened and withdrawn, this thin woman with sparse, stringy hair looked at us as if ready to weep, but tears did not come. As we were quick to appreciate, her dry eyes were a consequence of the disease: the early engorgement and later exhaustion of salivary and lacrimal glands known as Sjögren's syndrome, which arises in the course of scleroderma. Indeed, so hidebound was the skin about her lids, that the lacrimal glands could not be expressed. And her relatively youthful appearance was due to the shiny reflection of her collagenous mask, which had lost its normal wrinkles and fissures. We fitted the known stigmata of her disease to the patient before us. It turned out that she displayed the characteristic involvement of skin, lungs, and blood vessels that results when normal connective tissue—a pliant web of complex sugars, tensile proteins, and clear capillaries—is replaced by a carapace of scar tissue: scleroderma.

Despite extensive research, we know next to nothing about the cause and care of this sometimes progressively fatal malady. We are unable to stop its major assaults on vital organs. And although we have learned to manage the crises of breathing, vessel spasm, and kidney failure that mark its progression, the underlying problem of scleroderma is as unsolved today as it was thirty years ago, when I first learned about the disease in medical school. Monographs have been published, hundreds of research reports have flowed into our journals, and clinical subsets and related syndromes have been described. But while we are now able to discuss with greater precision the effects of scleroderma on the major organs, we remain as helpless as the doctors who forty years ago watched a network of collagen fibers tighten inexorably about the hands of Paul Klee.

As the weeks progressed, our patient responded to symptomatic treatment: vasodilators, artificial tears, hydration, and a soft, palatable diet. A caring staff managed to enter into a kind of communication with Mrs. Kahaner. This enterprise was made difficult by her persistent belief that she was bedded in the municipal hospital of Reykjavik, and that the Icelandic police were monitoring her words and behavior. This delusional system and associated fantasies were elicited by the house staff and consulting psychiatrists, who with patience and a dram or two of the newer psychotropic drugs dampened her fears of the Arctic constabulary. She was put at vague mental ease for the first time in months.

Nevertheless, the further management of her psychiatric troubles eclipsed her somatic complaints when the time came to plan her disposition after discharge. It was clear to the medical staff that she required close, periodic monitoring of her scleroderma, and equally clear that she belonged in some sort of institution that would shelter and clothe her, and where her madness would be managed in such a fashion that she could avoid the injuries of weather and a violent city. She needed what she had never found, an asylum—in the original sense of that word: sanctuary, from the Greek word *asylon,* meaning "inviolable."

However ignorant our profession may be of the proper management of scleroderma, it is safe to say that we know less of the disorders we call schizophrenia. Mrs. Kahaner, it turned out, had been in and out of various mental hospitals for at least two decades; she was divorced, and no relatives could be traced. Her recent confinements had lasted only a month or two; her prompt discharges were attributed to "remissions" induced (so her keepers believed) by various combinations of a broadening pharmacopoeia. After each discharge, she appeared with less regularity at follow-up clinics. For the past three years, interrupted only by two admissions to the emergency room of Roosevelt Hospital because she had been mugged, she divided her nights between the doorways of department stores and the rest rooms of Pennsylvania Station.

Mrs. Kahaner had joined that subclass of "deinstitution-alized" mental patients who have congregated in our cities in a kind of behavioral mockery of the consumer society. These shopping-bag ladies follow trails of private acquisi-tion not too dissimilar from those of their saner, middle-class sisters. From the avenues of the West Side to the doorway of Brooks Brothers, from the arcades of the sub-ways to the entrance of B. Altman's, the sad battalions of mad ladies course our streets, loaded with possessions tucked in tattered paper bags. And the bags still carry their mes-sage of fashionable competition: *Bloomingdale's! Bonwit Teller! Bergdorf Goodman!* The bags, in anarchic disorder, are in turn stashed in that other symbol of our abundant life: a shopping cart from the supermarket. There is probably no other country in which the madmen and madwomen so neatly exhibit the claims of local enterprise. I cannot recall similar public displays of private goods among the cat la-dies of Rome, the mendicants of Madrid, the down-and-out of London and Paris. Madness is surely as prevalent in those cities as in New York, but mockery of The Shopper in the person of the bag lady strikes me as pure Ameri-cana.

However, the disposition of poor Mrs. Kahaner after Bellevue will be governed by fashions in attitudes toward the mentally ill that know no borders. Motivated by the observation that the overt aberrancy of the badly disturbed can be managed by the wonders of psychopharmacology, our asylums have evolved over the past decades into homes of only temporary detainment. The therapeutic rescue fan-tasy—as current in Paris as in Glasgow, in London as in Albany—has been joined to the concept that the diagnos-tically insane will be more humanely treated, or achieve greater personal integration, if they are permitted freely to mix with the "community."

It has always seemed to me to constitute a fantastic notion that the social landscape of our large cities bears any direct relationship to that kind of stable, nurturing community which would support the fragile psyche of the mentally ill. Cast into an environment limited by the welfare hotel or park bench, lacking adequate outpatient services, prey to climatic extremes and urban criminals, the deinstitutionalized patients wind up as conscripts in an army of the homeless. Indeed, only this winter was the city of New York forced to open temporary shelters in church basements, armories, and lodging houses for thousands of half-frozen street dwellers. A psychiatrist of my acquaintance has summarized the experience of a generation in treating the mentally deranged: "In the nineteen-fifties, the mad people were warehoused in heated public hospitals with occasional access to trained professionals. In the sixties and seventies, they were released into the community and permitted to wander the streets without access to psychiatric care. In the eighties, we have made progress, however. When the mentally ill become too cold to wander the streets, we can warehouse them in heated church basements without supervision."

Clearly, what is lacking is the concept of asylum: a space where the mad and deranged can be protected not only from their internal demons but from the harsher brutalities of the street. Our loss of the asylum as an ideal can be traced to a number of social vectors, many of which take origin in the most altruistic of motives. Deinstitutionalization— a horrid word—became practical when the advent of palliative chemotherapy made possible the release of inmates whose overt behavior, on discharge, posed no immediate danger to patient or citizen. But despite the obvious fiscal and administrative relief afforded the state and local authorities, release would not have become widespread were it not for the vigorous efforts of well-meaning advocates of civil rights. These activists were rightfully convinced that commitment of the unwilling to mental institutions might represent the first step toward a Soviet-style incarceration for dissidents. Moreover, the return of the mad to the community was in accord not only with the rosy visions of fellowship and liberation that became prevalent in the sixties but also with a general distrust of arrangements made by any authority, be it juridical, governmental, or medical. And a prominent role was played in the attack upon the concept of asylum by radical sociologists, psychiatrists, and historians.

Absent any real knowledge of the nature of schizophrenia, one attractive path out of the forest of ignorance would be to deny that the term itself had meaning. It was not surprising, therefore, that the revisionist psychiatrist R. D. Laing seized the temper of the sixties and argued that those called "schizophrenic" were not sick in the medical sense. To Laing and his adherents, the schizophrenic— a soul more sensitive than the ordinary person— has made an entirely appropriate adjustment to an insane world. Since mental "disease" is only definable in relationship to a fractured, social framework of bourgeois values, *any* accommodation is acceptable, and the job of the healer is to adjust the discomfited to his level of accommodation. The Laingians argued that since a century of scientific inquiry had failed to prove that madness can be adequately described by the medical model, it was time to abandon the concept of schizophrenia as a disease. And if we can dispense with the therapeutic model, why not abandon those institutions that exist to treat this now nonexistent malady?

Most physicians, unlike other "mental health professionals," cannot be easily persuaded by this argument. For one reason, the argument suffers from what has been called the reification fallacy. This fallacy mistakes the *name* of a condition, or syndrome, for its *effects*. Were we to apply such flawed reasoning to the example of scleroderma, for instance, we would have to abandon the practice of making that diagnosis in Mrs. Kahaner's case. But by so doing, one would not alter in one sad, clinical detail the narrative of how our patient became trapped by collagen. Nor would her treatment, I daresay, be vastly improved by deinstitutionalizing Mrs. Kahaner before we had dilated her blood vessels, given her intravenous fluids, and helped her to breathe. In the example of scleroderma—a disease as enmeshed in mystery and as difficult to treat as schizophrenia—we are forced to resort to palliation, to support, and to clearheaded observation. I fail to see how our ignorance of the etiology of an illness gives us the right to assume that it is due to social maladjustment.

Indeed, the medical model may prove more effective. This model has already paid off in the case of lithium-sensitive, manic-depressive psychosis and in Alzheimer's disease. Until their basis in organic malfunction was identified, these conditions—like schizophrenia—were often attributed to social or familial wounds.

But the Laingian assault on the medical model and the modern response to that model—the asylum—has also been joined by the more urgent attack of historians and sociologists. They argue that mental hospitals subjugate and victimize the mad in the name of therapy, when what the mentally deranged require is liberation from the shame associated with the label of madness. Mental hospitals, so goes this charge, have since the seventeenth century served chiefly as schools for the brutal induction of shame; their abolition can only benefit the mad. This argument, in its most persuasive and scholarly form, is presented by the late Michel Foucault, the iconoclastic French historian of sexuality, prisons, and medicine.

In his major opus, *Madness and Civilization,* Foucault traces the history of mental institutions to the "Great Confinement" of the classical age. A royal edict of April 1656 led to the establishment of the Hôpital General, a series of Parisian hospitals for the confinement of mendicants, fools, and other idle folk of the street. These institutions, some of which were founded as centers for the sequestration of lepers in medieval times (lazar houses), became depositories for beggars, the unemployable, and the mentally feeble. In keeping with the new commercial spirit of the age, the Hôpital General and its sister institutions of England and Germany set the inmates to productive work and small manufacture in order to prevent, in the words of the royal edict, "mendicancy and idleness as sources of disorder."

Foucault describes how the confined populace replaced the leper as an excommunicated class:

> The asylum was substituted for the lazar house—the old rites of excommunication were revived, but in the world of production and commerce. It is in this space invented by a society which had derived an ethical transcendence from the law of work that madness would appear and soon expand until it had annexed them [*sic*]—the nineteenth cen-

tury would consent, would even insist that to the mad and to them alone be transferred these lands on which, a hundred and fifty years before, men had sought to pen the poor, the vagabond, the unemployed.

In these former lazar houses, the Age of Reason confined the socially undesirable: thief and demented, idler and aged, fool and madman. And so that strange mixture of the socially unwanted remained in chains or at forced labor until the advent of the French Revolution. It was at a branch of the Hôpital General, the Bicêtre, that Philippe Pinel struck these chains in an episode that signals the birth of liberal philanthropy. Pinel was confronted by an official of the Revolution, Georges Couthon, seeking suspects for trial. Pinel led him to the cells of the most seriously disturbed, where Citizen Couthon's attempts at interrogation were greeted with disjointed insults and loud obscenities.

It was useless to prolong the interviews. Couthon turned to Pinel and asked: "Now, Citizen, are you mad yourself to seek to unchain such beasts?" Pinel replied: "Citizen, I am convinced that these madmen are so intractable only because they have been deprived of air and liberty!"

"Well, do as you like with them, but I fear you may become victim of your own presumption." The official left, and Pinel removed the chains. The rest of the tale is the history of the modern mental hospital, at least until deinstitutionalization.

Now one would have thought that Foucault would come out cheering for Pinel. But no. In keeping with a tradition of radical criticism that includes the names of Herbert Marcuse and Ivan Illich, Foucault reserves his hardest blows for Pinel and the therapeutic reforms of bourgeois liberalism. Pinel stands accused by Foucault of substituting the psychological chains of the medical model for the iron shackles of the old lazar houses. The madman had become trapped in the birth of the asylum, and this new, therapeutic community confronted him with a new menace in the history of madness: guilt.

> The asylum no longer punished the madman's guilt, it is true, but it did more, it organized it for the man of reason as an awareness of the Other, a therapeutic intervention in the madman's existence.

Foucault, displaying that curious affinity of advanced French thought for the punitive tableaux of the Marquis de Sade, goes on to grieve for the unchained patients of the nineteenth century. He poses a very fashionable paradox:

> The dungeons, the chains, the continual spectacle, the sarcasms were, to the sufferer in his delirium, the very element of his liberation. . . . But the chains that fell, the indifference and silence of all those around him confined him in the limited use of an empty liberty. . . . Henceforth, more genuinely confined than he could have been in a dungeon and chains, a prisoner of nothing but himself, the sufferer was caught in relation to himself that was of the order of transgression, and in a non-relation to others that was of the same order of shame. . . . Delivered from his chains, he is now chained, by silence, to transgression and to shame.

Foucault's analysis of the relationship between keeper and inmate, between authority and the governed, extends to the interaction between doctor and patient. He points out that life in the asylum constituted a microcosm in which the values of a bourgeois state were symbolized: relationships between family and child, centering on authority; between transgression and punishment, centering on immediate justice; and between madness and disorder, centering on the social contract. Since the doctor stood *in loco parentis* for the whole society, Foucault suggested:

> It is from these [relationships] that the physician derives the power to cure, and it is to the degree that the patient finds himself, by so many old links, already alienated in the doctor, within *le couple médecin-malade,* that the doctor has the power to cure him.

This strikes me as a remarkable passage. To begin with, one can note that the French expression *le couple médecin-malade* has the same general meaning as the English "doctor-patient relationship," but in French it acquires a sexual overtone in its reference to the marriage couple. Secondly, Foucault adds the trendy spice of "alienation" to the principles of therapy: the madman, already alienated from the old bourgeois game of power and property, recapitulates that experience with his doctor. Finally, Foucault seems to equate alienation with the concept of transference. In psychoanalysis, transference—that strong bond between patient and therapist—is a positive element in self-discovery and possible cure. The only bond identified by Foucault is that of alienation, and this gives the doctor, as *locum tenens* of an alienating system, the "power to cure." But for Foucault, a cure is only a partial blessing, since it implies loss of freedom. Foucault has thus shifted the therapeutic setting from the Freudian one-on-one interview to the parade ground of the asylum, where the doctor as drillmaster instructs alienated conscripts in rules of the barracks.

In all aspects of his criticism of the asylum, Foucault uses as his point of reference a golden age of madness, which for him was the medieval age, where fools and crazies formed an integral part of the unregulated life of the street. Here, Foucault's vision of integration merges with that of the antiauthoritarian Left of the sixties. Foucault and the civil libertarians view the ameliorationist attitudes of mental health professionals as a destructive force in the battle for self-realization. They consider prisons, asylums, hospitals, and their squadrons of social workers, psychiatrists, and psychologists as the elements of a police state designed to censor the self-expression of the mad.

In the golden age of Foucault's medieval city, fools and madmen added to the richness of everyday life by their unique insights and startling behavior. The "reforms" of the Age of Reason destroyed this organic fabric and turned it into a straitjacket. That is the charge of Foucault, heard by the intellectuals of the West as the asylums have been emptied, mental health budgets cut, and the church basements filled.

But, I should have thought, there is little chance that poor, mad people find in our society even the glint of a golden age. In the cities of America, where the Mrs. Kahaners wander outside asylums in solitary danger, where violence is unchecked, and where the aggressive roam in quest of drugs and easy victims, we have, instead, partly reverted to the Hobbesian state of nature. I am frightened by that state and frightened for the army of fools and madmen we have let loose in it. To counter this condition, I, too, have a vision of a golden age, which by no means corresponds to the present-day treatment of the mentally ill. For me, it is the dream of the Age of Reason as articulated in the liberal philanthropy of Pinel. He gave us his vision of a true asylum for those shocked by the wars of the mind, an asylum that is, perhaps, more truly civil than that which Foucault presents. Pinel, in his 1801 treatise on the nature of madness, describes the hospital at Saragossa, where there was established

> a sort of counterpoise to the mind's extravagances by the charm inspired by cultivation of the fields, by the natural instinct that leads man to sow the earth and thus to satisfy his needs by the fruits of his labors. From morning on, you can see them—leaving gaily for the various parts of a vast enclosure . . . sharing with a sort of emulation the tasks appropriate to the seasons, cultivating wheat, vegetables, concerned in turn with the harvest, with trellises, with the vintage, with olive picking and finding in the evening, in their solitary asylum, calm and quiet sleep. The most constant experience has indicated, in this hospital, that this is the surest and most efficacious way to restore man to reason.

When we next meet Mrs. Kahaner, now lost in the shuffle between discharge from Bellevue and her unkept appointments at Creedmore, huddled in the closed doorways of Lord & Taylor, we might well ask whether her fate in our fractured city is better guided by the philanthropic vision of Pinel or the trendy critique of his dream by Foucault. Until reason and science unlock the shackles of her illness, we need, I believe, to give rest to this bag lady in a sanctuary, a therapeutic community—an asylum.

Cholera at the Harvey

ON the third Thursday evening of every month from September to May an auditorium in New York is filled with physicians, scientists, and students who await a lecture sponsored by the Harvey Society. Promptly at eight, the speaker, his guests, and officers of the Society, having wined and dined, wend their way to the front rows; their evening dress contrasts with the mufti and jeans of the contemporary audience. Both dress and lecture are formal, in deference to a tradition dating from the founding of the Society in 1905. Fearing that the medical profession of the city would not respond in any great number to the infusion of science into the world of practice, the founders bravely agreed to "wear our evening clothes, sit in the front row, and show the speaker we appreciate him." Happily, this elitist approach did little to dampen the brisk enthusiasm with which this series of lectures has been greeted over the following eighty years. In consequence, the annual volumes of Harvey

lectures document the dazzling transformation of medical
science in our century: of genetics from laboratory artifact
to molecular engineering; of physiology from muscle twitches
to the frontiers of crystallography; of cell biology from the
first glimpses of organelles to the frozen faces of membrane
lipids. Each month, with astonishing regularity, our lec-
turers—Nobel laureates and academicians, dons of the clinic
and jugglers of the gene—display yet another jewel of discovery mined from the unpromising rock of Nature. They
have amply fulfilled the charge of the society:

> . . . to foster the diffusion of scientific knowledge in
> selected chapters of the biological sciences and related areas
> of knowledge through the medium of public delivery and
> printed publication of lectures by men and women who are
> workers in the subjects presented, and to promote the de-
> velopment of these sciences.

Recently, the Harvey lecturer was Martha Vaughan of
the National Institutes of Health. Her topic was "Choleragen, Adenylate Cyclase and ADP-Ribosylation"—a title that
yields little promise of narrative splendor to the casual
auditor. The subject, however, was cholera, the disease of
lethal history, of wanton epidemics, of universal fear. It is
safe to say that cholera—before its virtual disappearance from
the developed world in the past century—ranked with bu-
bonic plague and pandemic tuberculosis in its capacity to
kindle the terror of an age. Yet, last month, as the lecture
unfolded, the disease itself remained as remote from the
audience as those cholera-struck villages in Bangladesh where
appropriate sanitation and modern treatment are unavail-
able. Here, in the plush dome of the Caspary Auditorium

at Rockefeller University, we heard, instead, of a neat solution to a molecular puzzle which had been locked in microbial genes since the beginnings of time.

It is now a hundred years since Robert Koch isolated the "Kommabacillus," or *Vibrio cholerae.* He had discovered *what* had caused the disease but did not know *how* the vibrio produced its assault on the gut. Perhaps because cholera had lost some of its sting in the Western world around the time of Koch's discovery, perhaps because the agenda of infectious diseases was too full, the problem of *how* did not approach the threshold of solution until 1959. Working in the service of a subcontinent still in the grip of cholera, two Indian investigators, De and Dutta, demonstrated that *V. cholerae* elaborated a crude, soluble factor that produced the disease in experimental animals. New York made its contribution when a graduate of its High School of Science, Richard Finkelstein (later of Dallas and Columbia, Missouri), collaborated with Dutta and isolated the toxin from the crude factor, which they diffidently called *choleragen.* This material—which proved to be the cholera toxin itself—was purified and shown to be capable, upon instillation in the gut, of provoking intense watery diarrhea in animals and in one brave human volunteer.

Research proceeded rapidly in many laboratories. It was established that the toxin is composed of two subunits, A and B. The B subunit has no direct toxic effects itself; it simply serves to target choleragen to a ubiquitous constituent of cell membranes. It is the A subunit, however, which is the business end of the toxin. When the A subunit gains access to the cell interior, it permanently switches on a key enzyme in the membrane of host cells: adenylate cyclase.

Since any agent that switches this enzyme on provokes the cells of our gut to pour out watery fluid (the stuff of cholera), it was unfortunate for man that the microbe had learned to turn this switch so efficiently. A remarkable set of findings, indeed! The intimate details of this story had been discovered in large part by our Harvey lecturer.

In sum, what emerged from this series of intricate biochemical and genetic studies was that cholera results from interactions between chemicals derived from both man and microbe. The patient is an obligatory contributor to his own symptoms. More remarkable still is the finding that similar mechanisms may be responsible for many aspects of our susceptibility to disease-causing microbes (diptheria, for example).

Listening to this tale of toxin and cell, it was difficult to remember that it had begun in a brave effort to find the cause of a common, virulent disease. But, a century ago, when Koch isolated the microbe from the "rice-water" stool of cholera victims, the disease was so public a menace that it became the stuff of myth and literature. The cholera epidemics of the nineteenth century, which ravaged the Old and the New World alike, left their mark equally on Naples and Boston, on Marseille and New York. Jean Giono, describing the 1838 epidemic in Provence, struck this sharp image of its human aspect:

> The cholera victim no longer has a face, he has a *facies*—a facies that *could only mean cholera*. The eye, sunk deep in its socket and seemingly atrophied, is surrounded by a livid circle and is half covered by the upper eyelid. These eyes will never have tears again. The lashes, the lids are impregnated with a dark, grayish matter. The cheeks have

43

lost their flesh, the mouth is half open, the lips glued to the teeth. The breath passing through the narrow dental arcades becomes loud.

The chill, first felt in the feet, knees and hands, tends to invade the whole body. Nose, cheeks, ears are frozen. The breath is cold, the pulse slow, extremely weak . . . The invalid is in an extreme state of agitation. He tries to rid himself of every covering, complains of unbearable heat, feels thirsty; forgetting all modesty, he flings himself out of bed or furiously uncovers his sexual parts. The pulse becomes more and more faint . . . The extremities take on a bluish tinge. The nose, ears, fingers suffer cyanosis; similar patches appear on the body.

[At the end,] the eyes are turned up, their brilliance has vanished, the cornea itself has thickened. The gaping mouth reveals a thick tongue covered with ulcers. The chest no longer rises. A few sighs. It is over.

[And in the charnel houses of the disease,] an almost ludicrous domesticity was mingled with the terrifying appearances of the curse which preceded time. The corpses continued to relieve themselves into shrouds now made of any odds and ends, old window curtains, sofa covers, tablecloths and even, in wealthy houses, bath covers. Chamber pots full to the brim had been placed on the dining room table, and people had gone to fill dishes, washbasins—and even flower pots—with that mossy, green and purple fluid that smelled terribly of the wrath of God.

I had unearthed this passage from Giono on the day after the Harvey lecture, prompted by ruminations which had begun in a taxi on the way home through empty streets leading from the East River. I recalled that only a mile or so downstream from where we had attended Martha Vaughan's careful narrative of twentieth-century science, cholera hospitals and quarantine shacks had been hastily erected to house victims of epidemics in the past century. And over the same route I traveled that night, tumbrels filled with corpses in filthy shrouds had been carted to distant burial pits.

Three waves of cholera had struck New York, in 1832, 1849, and 1866. In those days the disease arrived by ship, carried in holds crowded by immigrants to the New World; it also came by land, carried by the unemployed from Montreal, Boston, and Philadelphia. By July of the first cholera year, 1832, 3,000 were stricken and 1,800 had died, of a population of 300,000. Everyday life was disrupted, citizens were afraid to walk the streets, a strict quarantine was established in the port and at the Canadian border. In the absence of any real understanding of the cause of cholera, or its mode of spread, the propertied burghers, by means of broadsides and newspapers, cast blame on the poor and immigrant classes for their "imprudent and intemperate ways." The authorities were urged to keep the poor segregated in their districts and to enforce temperance.

But the city fathers did more than encourage abstinence and obedience to divine will; they also disbursed $118,000 with which to build temporary cholera hospitals and to clean the streets. Predictably little could be done to ameliorate the disease in hospitals, where the afflicted encountered overworked doctors and useless remedies. The chief utility of hospitals was to segregate the sick from the fit. Sanitary measures, devised to insure the disposal of refuse, succeeded mainly in keeping the wealthier quarters of the city free from offal. That first summer 3,000 died; not until winter fell did the epidemic wane.

Debates in the popular press and medical journals expressed a clear split of opinion as to how recurrence might be prevented. On the one hand, it was urged that tough quarantine and detention measures were the only solution; immigration should be halted, the newly infected must be identified, and all suspected cases were to be strictly isolated from the general citizenry. On the other hand, more progressive opinion held that the disease was spread by water that had been contaminated by sewage. An attack on this problem—in a city of outhouses—would involve massive outlays of money to provide a reliable source of fresh water. In the event, both courses of action were taken: immigrants were rigidly screened and quarantined, and plans for a new water supply were drafted. By 1842, a major project was undertaken. The Croton resevoir system, which still brings New York its water, was begun at an ultimate cost to the city of more than $11 million in the next two decades—a sum many multiples of the city's yearly health and sanitation budget!

The epidemic struck again in the wake of a new wave of immigration, which followed a year of European revolution. By December of 1848, cholera was rampant in Le Havre. By spring, alarmed at new cases in New York, the authorities appropriated $55,000 to "give our city a cleaning as it has not had in years." This sprucing up did not prevent cholera, nor did the tough quarantine laws and inspections at points of immigration. In the summer of 1849, 18,000 fell ill and 8,000 died, of a population that had risen to over half a million. Despite the efforts of President Zachary Taylor, who declared a national day of fasting and prayer, cholera killed 1,178 on August 11; this was the high-water mark of the disease in New York. New hospitals were built and teams of sanitary inspectors routed those suspected of the disease from tenements. Ships filled with stricken immigrants were kept from docking and consequently crowded the harbor. Shops closed, the streets emptied once more, and public gatherings were canceled. The epidemic ran its awful course, nevertheless; only the advent of the winter months permitted the city to return to its normal patterns.

By 1865, the sanitary movement had spread its enlightened wings. The U.S. Sanitary Commissioner, Templeton Strong, responded to news of European cholera with the assertion, "Sanitation, not quarantine, is the answer!" A strong health bill established the New York City Board of Health; the Croton reservoir system was partly in place. Yet cholera struck again in 1866. This time, sanitary measures and isolation procedures were both set in motion. It was now clear to medical experts that fecal contamination of the water supply had much to do with the spread of the disease. The aqueducts to Croton and the rudimentary sewage systems were rigorously cleaned. Elisha Harris and Stephen Smith, of the Board of Health, established sixty subdistricts in each of which teams of officials examined all possible causes. Suspects were trundled off to several temporary hospitals on land and to newly established floating hospitals in the harbor. Vigorous efforts were launched to displace over 115,000 cellar dwellers from crowded hovels and to house them in cleaner dormitories. *The New York Times* was sufficiently reassured by this mixture of sanitation and isolation to assure its affluent readers that "thus guarded, there is not much danger of disease ranging in the more comfortable quarters of the city." The Board of Health spent over $250,000 that summer. Its two-pronged attack on the disease may have been responsible for the fact that in a city grown to 750,000, only 1,200 contracted cholera, and only 600 died.

Sanitation improved remarkably in the next decade, and the temporary hospitals were closed; Ellis Island served as a screen against importation of contagious disease. Thus, by 1892, when cholera was rampant in Russia and southern Europe and had spilled over to Montreal and other East Coast ports, New York escaped for the first time. The city, having refused to pour all of its resources into either the sanitary or the quarantine approach, had successfully played the field, so to speak. New York had won its fight with cholera forever—even without a clear understanding of its etiology on the part of doctors—by the remarkable expedient of throwing a great deal of money at a public problem.

My reading of this history, as I've indicated, was stimulated on the journey home from the splendid auditorium of Rockefeller University. The streets of the comfortable East Side were empty at 10:30 P.M.; shutters and iron gates protected the boutiques of Madison Avenue. What a contrast between this barren cityscape and the bustling street life remembered from barely three decades ago, when one wandered safely through this part of Manhattan at all times of the day or night! The empty streets sent the message that we are living in the midst of a twentieth-century epidemic: the epidemic of violent crime.

It is a sad commentary on this epidemic that, in 1967, our Society was forced to shift the site of the Harvey Lectures from the imposing Romanesque building of the New York Academy of Medicine, at 103rd Street and Fifth Avenue, to the less hostile environment of Rockefeller University, at 66th Street and York Avenue. In the mid-sixties, attendance had begun to drop owing chiefly to fear on the part of the audience of muggings, car vandalism, and the squads of angry teenagers which invaded the avenue from the barrios and Harlem. In consequence, like our predecessors who avoided the streets in deference to cholera, we, too, have beaten a retreat to the perimeter of "more comfortable quarters of the city," the streets of which are now largely empty at night. Attendance perked up again promptly once we moved downtown; the doctors' cars now remain intact.

Now, it may be stretching an analogy to compare epidemics of cholera—caused by a known agent—with that epidemic of violent crime which is destroying our cities. It is unlikely that our social problems can be traced to a single, clearly defined cause in the sense that a bacterial disease is "caused" by a microbe. But, I daresay, social science is about as advanced in the late twentieth century as bacteriological science was in the mid-nineteenth century. Our forerunners knew *something* about cholera; they sensed that its spread was associated with misdirected sewage, filth, and the influx of alien poor into crowded, urban tenements. And we know *something* about street crime; nowhere has it been reported that a member of the New York Stock Exchange has robbed a poor, black teenager at the point of a gun. Indeed, I am naively confident that an enlightened social scientist of the next century will be able to point out that we had available to us at least some of the clues to the cause of urban crime. After all, the comparative epidemiology of violent street crime does tell us that Swedes and Japanese tend to kill themselves more frequently than they kill other people and that the Swiss mainly kill time.

It is quite likely that the chronic warfare between rich

and poor, black and white, plays *some* role in our particular epidemic. Armchair sociologists of the Left suggest that the collective psyche of our nation is flawed by a "streak of violence" that extends from our dispossession of Native Americans to the mayhem of film and television and that this flaw is reified in our infatuation with guns. Bigots of the Right insist that violent crime is chiefly a social disease brought about by our lax morality, the decline of fundamental religion, and our failure either to keep blacks in their place or to limit immigration from the Caribbean.

Based on these opposing etiologic hypotheses, two quite different strategies have been formulated to deal with our new epidemic of violence. On the one hand, ameliorationists propose that major outlays of public funds are required to root out the basic inequalities of our entrepreneurial system: Win the war against poverty! Grant true social justice to blacks and Hispanics! And violence—the symptom of inequality—will abate. It does sound a bit like the sanitary approach to cholera. The realists doubt that any outlay of money will suffice to eliminate the roots of violence. They urge us, instead, to beef up the police, to impose swift and Draconian punishment, to build more prisons, and to reward murder with certain death—all whilst staunching the influx of the potentially violent from without our borders. Their answer to our epidemic closely resembles the quarantine approach to cholera.

Well, to my mind, we will probably have to make a major effort in both directions. No amount of money devoted to Head Start programs today will deter the tough punk who has stabbed a nun. But it is also probably useless to assume that there will ever be enough police, sufficiently quick judgment, or enough prisons to cope with the violent eruptions of the unemployed underclass that fills our cities. Nor is it likely that we can detain the potentially violent—or quarantine the roving gangs—not unless we wish to live with an entirely different legal code, indeed. No, I'm afraid we are faced with the same predicament that confronted our predecessors when they were coping with cholera. If we're at all serious, if we're really afraid, and if we also want social justice, we shall have to throw a great deal of money at our epidemic and head off in two directions at once. We need both school lunches *and* more cops, Head Start programs *and* tough law enforcement, bilingual education *and* more jails. Absent exact knowledge, we will probably have to make do with a series of untested notions and try even unproved remedies. It will not be a neat solution, but it will be the kind of solution that worked once before, here in New York, when the streets were empty and the city decided to do something about it.

In Quest of Fleck: Science From the Holocaust

In 1947, there appeared in the *Texas Reports on Biology and Medicine* (9:697–708) a report by Ludwik Fleck on "Specific Antigenic Substances in the Urine of Typhus Patients." Its opening sentences are probably unparalled in the annals of biological science:

> The search for specific antigenic substance in the urine of typhus patients was initiated in Lwow, 1942, under German occupation. The original plan was to elaborate a test giving earlier diagnosis than the Weil-Felix reaction. . . . In addition to elaborating a diagnostic test, it was also thought to utilize the urine of typhus patients as a source of specific antigens for the preparation of a preventive vaccine, very urgently needed at this time.
>
> In May 1942, the results were reported at a staff meeting of the "ghetto" hospital in Lwow, and, several months later, the author was deprived of his collaborators who were destroyed by the Germans.

The report goes on to describe the efficacy of this vaccine in typhus fever, the second-leading cause of death in the concentration camps.

> The author, his collaborators, and 32 volunteers were vaccinated . . . Later, 500 people in a concentration camp at Lwow were vaccinated. With few exceptions statistics of the vaccinated were unfortunately lost. Records are available only in regard to the author, his family, and two other persons, all of whom contracted typhus and recovered after a mild or abortive course of the disease. A large number of the vaccinated in the camp did not contract typhus although they were exposed to typhus infections. In contrast, the majority of the nonvaccinated prisoners contracted typhus with a fatality rate of 30%.

Now, neither the field of study described in this report (the serology of typhus) nor the journal in which it is reported would ordinarily engage my attention. However, the sequence of events that led to my encounter with this unusual medical scientist seems worth recounting, because it reflects the tides of discovery and rediscovery, the charting of which, ironically, constitutes Fleck's chief legacy to the history of science.

The development by Ludwik Fleck of a practical vaccine against louse-borne typhus from the urine of the afflicted was directly responsible for his survival in the camps. The Germans, as anxious to protect their troops as they were indifferent to the fate of their captives, forced Fleck first to produce this vaccine in the ghetto hospital of Lwow (formerly called Lemberg) and later in the camp "hospitals" of Auschwitz and Buchenwald. But in the course of his studies of the serological response to his vaccine, he made a

53

signal scientific discovery, one that directly illuminates my own field of research, which is the role of white blood cells in inflammation. Using as his tools nothing but the clinical microscope, a few common laboratory dyes, and the (unfortunately) abundant peripheral blood of patients with typhus, he described the phenomenon he called "leukergy."

The first description of leukergy also appears in the same issue of *Texas Reports* in an article by Fleck and Z. Murczynska (the issue was devoted to reports of medical research caried out in wartime Poland). The authors present evidence that the white blood cells (neutrophils) in the blood of patients with typhus clumped into tight little clusters after a few minutes of incubation at 37 degrees Centigrade. Fleck went on to determine that these white cell clumps were also observed in the blood of rabbits that had been experimentally infected with certain bacteria or their endotoxins and, indeed, in the course of many infectious diseases. This response of the white cells, he concluded, prepared them for such other functions as sticking to the walls of small blood vessels, squiggling out of the vessels toward offending microbes, and engulfment of the bugs.

The description and analysis of leukergy, first encountered in the unspeakable setting of the death camps, occupied Fleck and his collaborators throughout the postwar period. Publications resulting from this work appeared not only in the Polish literature and in translated form in the *Texas Reports,* but also in more widely read journals, such as the *Schweizerische Medizinische Wochenschrift* and *Acta Haematologica* (Basel), in *Le Sang* (Paris), and in the *Archives of Pathology* (Chicago). In 1949, Fleck was awarded the scientific prize of the city of Lublin for his discovery of leukergy, which he classified as an important biological response of white blood cells to injury or infection. It is therefore remarkable that this work had no impact whatsoever upon the scientific community at large. Indeed, a search of the *Science Citation Index* (Philadelphia) for the years 1965 to 1980 reveals fewer than half a dozen references to this discovery.

But leukergy is very much alive, although living under assumed names. In the past decade, studies from many laboratories, including my own, have shown that the "aggregation" and the "adherence" of neutrophils are among their very early responses to a variety of inflammatory insults. The study of the function of white blood cells in disease is now an enterprise that must engage several thousand investigators throughout the world. This group, or "thought collective" (to use Fleck's phrase), has come to appreciate that the altered surface properties of neutrophils, their aggregation in blood vessels—as reflected by their clumping in the dish—probably accounts for some of the more lurid complications of allergy, infection, and shock. Hundreds of articles, scores of learned reviews, and extensive book chapters document this role—all without mention of Fleck. The trial vessel of leukergy has sunk without trace.

If this be the fate of Fleck's major experimental opus—his work on typhus vaccine has not only been superseded but rendered superfluous by advances in hygiene, new pesticides, and the advent of antibiotics—how did the phenomenon of leukergy emerge from the depths of obscurity? The answer lies in the revival of Fleck in another context: as a philosopher and historian of science, as a forerunner of Thomas Kuhn, and as an important thinker in the sociol-

ogy of science. In 1935, Fleck had published a book entitled *Entstehung und Entwicklung einer wissenschaftlichen Tat-
sache: Einführung in die Lehre vom Denkstil und Denkkollektiv
(Genesis and Development of a Scientific Fact: Introduction to the
Study of Thought Style and the Thought Collective)*. The book,
written by a Jew, was unpublishable in Germany and
therefore was published in Switzerland. Only 640 copies
were printed: 200 were sold, and of these one found its
way into the library of Harvard. While still a member of
the Harvard Society of Fellows, the illustrious philosopher
and historian of science, Thomas Kuhn, came across Fleck's
book, stimulated by a footnote in a work of Hans Reichen-
bach. Kuhn, who refers to Fleck in his *Structure of Scientific
Revolutions,* recollects: "In twenty-six years I have encountered only two people who had read the book indepen-
dently of my intervention."

It was at a meeting of another Society of Fellows, that
of New York University, of which Kuhn was also a transient member, that I first heard mention of Fleck the philosopher. Consequently, when a new edition of Fleck's book
appeared, with a foreword by Kuhn, well edited and an-
notated by Thaddeus J. Trenn and Robert K. Merton, I
was eager to grapple with it. The book was challenging on
three levels. First, Fleck presents an original analysis of how
scientific facts, especially those of biology, are necessarily
contingent upon the social context in which they are established. A discovery is both the product of, and a factor in,
a discrete social setting. Second, a brief biographical sketch
provided by the editors describes the harrowing circumstances under which the author's major discoveries were
made: a tale of triumph plucked from horror. And finally,
a footnote in the book, which refers to Fleck's work on

leukergy, rings the bell of recognition for someone who has spent the better part of his scientific career worrying about white cells. The book deserves our first attention.

Genesis and Development of a Scientific Fact argues that scientific "facts" are not absolute, but relative, and that they necessarily relate not only to the general social scene but especially to the "thought styles," or modes of perception of those individuals who compose the "thought collective" in a particular field of inquiry. Fleck points out that the so-called Vienna Circle of epistemologists, such as Moritz Schlick and Rudolf Carnap, who were trained in the physical sciences, considered that thought processes (cognition) were *fixed* and absolute, a view not unlike that of the new structuralists. On the other hand, they considered empirical facts to be *relative*. In contrast, the philosophers with a humanistic, or sociological, background (Emile Durkheim, Lucien Levy-Bruhl) considered scientific facts to be fixed, whereas, human *thought* was relative. Fleck asks:

> Would it not be possible to manage entirely without something fixed? Both thinking and facts are changeable, if only because changes in thinking manifest themselves in changed facts. Conversely, fundamentally new facts can be discovered only through new thinking.

Fleck went on to suggest that cognition itself is essentially a social activity, since the existing stock of knowledge exceeds the range available to any one individual. Knowledge is not generated *in vacuo* by a particular consciousness, or by any one person, but by a thought collective, which he defines as a

> community of persons mutually exchanging ideas or maintaining intellectual interactions . . . It also provides the

special "carrier" for the historical development of any field of thought, as well as for the given stock of knowledge and level of culture (thought style).

In a telling analogy, he recapitulates this theme:

> If an individual may be compared to a soccer player and the thought collective to the soccer team trained for co-operation, then cognition would be the progress of the game. Can an adequate report of this progress be made by examining the individual kicks one by one? The whole game would lose its meaning completely.

This mixture of epistemology and social theory is stirred by means of a detailed analysis of a scientific "fact." Fleck, a trained serologist, analyzed the history of the Wassermann reaction, the first blood test for syphilis. This test was his "fact," and he traced its roots to earlier notions of impure blood. Fleck compared what August von Wassermann and Carl Bruck wrote about their discovery years after their first observation with what they had originally described.

In their first two papers on the subject, Wassermann and coworkers had described use of a serological method (the complement fixation reaction) to identify *antigen* in watery extracts of syphilitic tissues. Antigens, of course, were supposed to signal the presence of the disease-causing microorganism. Positive results were obtained in sixty-four of seventy-six extracts. An *antibody,* that which is detected by the test as now performed, was found in only forty-nine of 257 samples of blood from syphilitics. Moreover, it was soon determined that the *antigen,* later identified as a ubiquitous fatty material called *cardiolipin,* was not only found in diseased, but also present in healthy, tissues. However, thanks to refinements in the technique introduced by Julius Citron and others (they used alcohol or acetone to make antigen), the test for *antibody* became positive in up to 90 percent of syphilitics. This is the "fact" of the Wassermann reaction.

Fleck quotes Wassermann's hindsight fifteen years after the fact: "I proceeded from the idea, and with the clear intention of finding a diagnostically usable amboceptor (antibody)." And after comparing this statement with the original work, Fleck concludes:

> The ultimate outcome of this research thus differed considerably from that intended. But after 15 years an identification between results and intentions had taken place in Wassermann's thinking . . . From false assumptions and irreproducible initial experiments an important discovery has resulted after many errors and detours. The principal actors in the drama cannot tell us how it happened, for they rationalize and idealize the development.

Recent accounts of such major discoveries as the helical structure of DNA or the discovery of the hepatitis-associated antigen would tend to confirm Fleck. But how were these "false assumptions and irreproducible initial experiments" transformed into an "important discovery"? Fleck is clear:

> It was the prevailing social attitude that created the more concentrated thought collective which, through continuous cooperation and mutual interaction among the members, achieved the collective experience and the perfection of the reaction in communal anonymity.

> Laboratory practice alone readily explains why alcohol and later acetone should have been tried besides water for

extract preparation, and why healthy organs should have been used besides syphilitic ones. Many workers carried out these experiments almost simultaneously, but the actual authorship is due to the collective, the practice of cooperation and teamwork.

Further description of the richness of this book would require much more extensive documentation. Suffice it to say that Fleck gives the best account, by far, of how the products of bench science become translated into journal articles and then integrated into textbooks. Only in texts, removed from the realities of quotidian work, do "facts" exist as such—and even these textbook facts have finite half-lives.

I would, however, like to dwell on one passage which deals with the main problem inherent in Fleck's view of facts as servants of collective fashion. What is the role of the original observer, the adventurer, whose sudden insights anticipate the consensus of the thought collective? Here we may find the judgment of Fleck, the sociologist, on Fleck, the philosopher, and Fleck, the biologist:

> . . . Such scientific exploits can prevail only if they have a seminal effect by being performed at a time when the social conditions are right . . . Had Vesalius lived in the twelfth or thirteenth century he would have made no impact . . . The futility of work that is isolated from the spirit of the age is shown strikingly in the case of . . . Leonardo da Vinci, who nevertheless left no positive scientific achievement behind.

In one sense, the rediscovery of Fleck by the philosophers and historians—and what I am certain will turn out to be his rediscovery by the thought collective of the neutrophil world (at least if one reader has anything to do with it!)—is the validation of this passage and of Fleck's major thesis. The appropriate thought collectives within the social sciences and human pathology were not ready to incorporate Fleck's own contributions, not until Kuhn had brought about his own revolution, not until Merton had turned the glass of social analysis to the life of the laboratory, and not until we had learned how infections and endotoxins activate the neutrophil.

Although brief, Fleck's biography, as outlined in the Trenn-Merton volume, evokes completely the agony of a Jewish intellectual who was torn between Polish and Germanic cultural traditions, finally to be betrayed by both. Born in 1896 in Lemberg, he attended Austrian secondary schools. He received his medical degree from the University of Lwow in 1922, shortly after the town reverted to Poland. His postdoctoral training in serology was with Rudolf Weigl of Lwow and, later, at the University of Vienna, where Clemens von Pirquet and Friederich Kraus were major figures. In 1928, he became head of a government bacteriology laboratory in Lwow and appeared to have achieved security, until he was dismissed (in 1935) in consequence of one of the not uncommon anti-Semitic seizures of the Polish people. Between 1935 and 1939, he was forced to earn his living by establishing a private, diagnostic laboratory of microbiology. Throughout the thirties his research interests (which had always been directed toward typhus, the Wassermann reaction, and host defense mechanisms) became joined to a broader interest in the social and humanistic features of the scientific calling, and he published several papers on history and philosophy, both before and after his 1935 book.

When Stalin and Hitler divided Poland, Lwow became

Russian, and perhaps due to his progressive (or at least anti-Fascist) views, he was promptly made director of the city's microbiological laboratory, finally gaining an appointment to the medical school. When the Germans occupied Lwow in 1941, he withdrew, as prescribed, to the ghetto hospital, from where his first observations on both the typhus vaccine and leukergy originated. In 1942, having involuntarily ceded the methodology for producing typhus vaccine to the Germans, he was sent to Auschwitz, where his sisters and their families perished. In 1944, Fleck was transferred to Buchenwald—still in the camp "hospitals," still precipitating antigen from urine to immunize against typhus. He was liberated by the American army in 1945 and, on his return, was received with honor by the Polish authorities. From 1945 on, he rose in the ranks of the new Marie Sklodowska Curie University of Lublin to full professor; in 1952 he became Director of the Department of Microbiology and Immunology at the State Institute in Warsaw. Despite many honors, including election to his country's Academy of Science in 1954, he had always wished to emigrate to Israel, permission for which was granted in 1957. He was appointed head of the section of Experimental Pathology of the Israel Institute for Biological Research and appeared headed toward a new burst of scientific and belletristic activity when he was stricken with Hodgkin's disease. He died in June of 1961.

I cannot say with certainty what the ultimate role of this medical amateur will prove to be among the lions of the social sciences. I am certain, however, that in my own field, the rediscovery of Fleck will prove useful. He correctly perceived that the sticking of neutrophils to each other—brought about, for example, by bacterial toxins—is the first step in *Activierung des Leukocytären Apparates* (activation of the leukocytic apparatus). Activation of this apparatus permits the stimulated cell to increase its locomotion, its engulfment of bacteria, and its power to kill the microbes. It will not be difficult, utilizing the technical tools and nomenclature fashionable in our modern thought collective, to repeat, extend, and incorporate the observations made by Fleck and his coworkers on the blood of patients with typhus.

Perhaps Fleck would have been pleased that leukergy has again been perceived as a "fact." (The image comes to mind of the thought collective carrying their triumphant captain off the soccer field.) But the final irony is this: If Fleck is correct in his book, then his discoveries of the typhus antigen and of leukergy were, of necessity, born in the social crucible of the Holocaust. The bacteriologist of the ghetto hospital of Lwow would probably have been the first to wish that his discoveries were unnecessary.

AIDS and Heat

New diseases are taking their turn at notoriety in the press—as if they were punk rock stars or banana republics with an insurgency problem. A few years ago it seemed as if the resources of the medical profession were entirely devoted to the fight against Legionnaire's disease. This respiratory ailment turned out to be due to a new microbe found lurking in the cooling systems of hotels and hospitals. The next threat of the decade was toxic shock syndrome, a catastrophe caused in large part by the affinity of certain staphylococci for the crevices of a feminine hygiene device. A lethal epidemic of heart, lung, and connective tissue disease then humbled the culinary pride of Spain when rapeseed oil was accidentally mixed with olive oil. Paraquat became a household word after the government of the United States declared war on the lungs of marijuana smokers. The Feds not only provided this herbicide to our Mexican neighbors as part of an international drug control program, but, *mirabile dictu,* recently authorized use of paraquat in Georgia's Chattahoochie National Forest to halt the harvest of our domestic pot crop. And, finally, each day's newspaper seems to detail further subversion of our Atlantic bluefish by PCBs; a more immediate menace to New Bedford than the rumbling *guerrilleros* of Guatemala.

This seemingly random list of headline news from the medical front is united by the remarkable fact that none of the maladies described was part of the medical curriculum a short decade ago. Does this mean that we are suddenly up against a new army of hostile microbes or pollutants never before encountered? Or are we simply calling old illnesses by new names? A bit of both, I suspect. We know, for example, that *Legionella* accounted for several outbreaks of pneumonia that preceded isolation of the organism; for this information we can thank epidemiologists who squirreled away frozen serum samples in the hope that these could be used for future antibody studies. And toxic shock syndrome had certainly been described, albeit in sporadic form, before the jumbo tampon came along. These are examples of older diseases, newly risen to epidemic proportion. But there are other instances in which man has become newly susceptible to microbes which had previously confined their mischief to other species. One can cite tularemia, not recognized in man until 1904, although appreciated as a pathogen in wild animals; psittacosis, which we have only recently shared with pigeons and parakeets; and Lyme arthritis, acquired from ticks by way of pets.

These observations were prompted by a rereading of Hans Zinsser's *Rats, Lice and History*—a once widely popular book on the history of infective diseases—and by newspaper accounts of public reaction to the newest assault on our well-being: acquired immune deficiency syndrome, or AIDS.

This summer of 1983 (see Addendum, p. 77), the toll of AIDS victims rose to approach 2,000 and mortality from the disease neared the 50 percent mark. Appearing this August at the bedside of Peter Justice, a forty-year-old AIDS patient at the Cabrini Medical Center in New York, Margaret M. Heckler, the Secretary of Health and Human Services, assured a press conference that she would urge Congress to approve $40 million for research in the next year on AIDS—a sum twice that originally requested. She had found the extra money in the back pocket of her vast enterprise; it seems that $22 million originally committed to "new furniture and construction" for the Rural Development Fund would instead be diverted to AIDS research. Indeed, her department and spokesmen for the NIH have vowed that research into AIDS will be the government's "top priority" in next year's budget.

If we assume that the incidence of AIDS will double in the coming year to 4,000 cases, we can calculate that $10,000 of AIDS research money is about to be allotted for each of the stricken—a sum, which, if expended in similar proportions on cancer, heart disease, and other lethal maladies, would swamp the federal budget. If I can still remember the appropriate figures, research to help the two million sufferers from Alzheimer's disease should, by this line of reasoning, be funded to the tune of $20 billion per annum. And, since the Arthritis Foundation estimates its clientele of patients at 32 million, my own field of investigation could lay claim to $320 billion!

If research on AIDS is to be funded so far out of proportion to the actual numbers of its victims, one has to ask why this particular epidemic has aroused such unusual anxiety. I suspect that what troubles us so deeply about

AIDS is that the disease may be caused by a uniquely worrisome virus, one that might at any moment escape the somewhat limited populations now affected to be loosed among the public at large. Over 70 percent of cases have been described in promiscuous male homosexuals, the remainder among heterosexual drug abusers, Haitians, and hemophiliacs. The *fear* of AIDS, however, fed by lurid accounts on television, radio, and in the press, extends to a much wider audience. Taxi drivers, I am told, worry over handling coins given them on Christopher Street; garbage collectors and health workers demand special costumes to service the homes and bedsides of AIDS victims; and Manhattan matrons have found excuses for dismissing their Haitian domestics. Since, at present, there is little evidence to suggest that such casual encounters have led to direct transmission of the disease, we might hazard the suggestion that the large emotional component of the public's fear derives as much from the special social status of those presently afflicted as from any real and present danger.

A reading of Zinsser on the arrival of syphilis in Western Europe discloses a historical precedent for these fears and anxieties. Indirect evidence has traced introduction of the disease to Naples by way of the crewmen of Columbus or their sexual partners. The first European epidemic broke out in a "sudden, intense, and widespread manner" after the French troops of Charles VIII occupied Naples in 1495; it soon became rampant among the soldiers, their camp followers, and the underclasses of the city. Depending upon the nationality of the recorder, therefore, syphilis was immediately dubbed either the "French" or the "Neapolitan" disease. In its infancy, syphilis presented a much more virulent and immediately contagious clinical picture than in succeeding centuries: the stricken were covered with pustular and vesicular eruptions, with painful ulcerations that covered much of the body.

"The ulcerations," Zinsser reports, "covered the body from head to knees. Crusts formed, and the sick presented so dreadful an appearance that their companions abandoned them and even the lepers avoided them. Extensive losses of tissue in the nose, throat and mouth followed . . . and then came painful swellings of the bones, often involving the skull. The disease itself, or secondary infection, caused many deaths."

But in a short fifty years, the disease changed its clinical presentation. By 1546, Fracostorius was able to report in his *De Contagione* that the painful, ulcerative disease had yielded to a chronic, less devastating illness: gummata were more common, the lesions themselves were less painful, and the overall disease caused fewer deaths.

The history of infectious diseases presents many similar instances of amelioration over time in the absence of specific therapy. Leprosy, a dreadful and commonplace ill in the medieval period, became—for reasons that remain obscure—remarkably less crippling as its prevalence waned in early modern Europe (See "Foucault and the Bag Lady.")

The consequences of scarlet fever, which had decimated the children of the Industrial Revolution, were already diminished in severity by the end of the nineteenth century, and the form of typhus known as Brill's disease—which was found among Jewish immigrants of Polish or Russian origin in New York—bore only faint traces of its epidemic precursor in Eastern Europe.

These accounts of the natural history of infections before the antibiotic era serve as useful controls for the experiments of nature afforded by our recent epidemics. In the centuries before antibiotics, guided by the unerring prejudice of folk wisdom, the citizenry blamed one or another minority for the outbreak and propagation of contagious disease: the beggar, the Jew, the gypsy, the *other*. Happily, in our time, the rapid advance of medical science has in good measure made this particular sort of scapegoating superfluous. Thanks to antibiotics, we will never know whether untreated Legionnaire's disease would have waxed and waned in severity over the course of a century; for the same reason, we have not made social outcasts of Pennsylvania's American Legionnaires. We neither fret over dangers posed to our communal health by the burghers of Lyme, Connecticut, nor scold the victims of toxic shock for their habits of personal hygiene. Only the gays with AIDS remain in double jeopardy of disease and ostracism.

The life history of our new infectious diseases has been crowded into a few busy years in which the first clinical description was rapidly followed by the discovery of an offending microbe and the design of appropriate chemotherapy. Their conquest was made possible by lessons learned in the campaigns against far older contagions, as tactics which had proved successful against plague, typhus, and tuberculosis were applied to tularemia, psittacosis, Legionnaire's disease, and Lyme arthritis. Such tactics were based on a long and painfully acquired body of basic and clinical knowledge, and were appreciably aided by the growth of industrial pharmacology. The cost of acquiring that body of knowledge—of microbiology, epidemiology, immunology, and pharmacology—has amounted over the years to many millions of dollars. I would wager that, were it possible to do a cost accounting of the sums expended per annum on work which directly or *indirectly* resulted in the successful treatment of the newer infectious maladies such as Legionnaire's disease, we might very well come up with a figure that would come close to $10,000 per victim.

Sustained by this line of reasoning, I support with enthusiasm Secretary Heckler's request for AIDS research money. Much of the $40 million will undoubtedly go to those laboratories of basic and clinical research which have, for other reasons, established the ground rules of modern immunology, according to which the search for a cause of AIDS will be conducted. We have already learned about AIDS by learning the principles of immunologic regulation, such as the functions of T- and B-cells and the role of helper and suppressor T-cell subsets. And our teachers have been drawn from laboratories, the funding of which was justified to Congress on the grounds that we would be better able to combat tuberculosis, rheumatoid arthritis, and cancer. We have learned about the genetic and biochemical bases of immunodeficiency from other laboratories, the funding of which was justified by appeals that we might be able to help the few children whose lymphocytes lack the enzyme adenosine deaminase, or a much larger group of patients who suffer from gout. It is therefore abundantly clear that research into any new disease such as AIDS not only draws upon, but seems likely to contribute to, the common pool of useful knowledge. Thus, our inquiry into AIDS can rely on a substantial body of clinical and basic science that is already in place and for which we have already paid. In the absence of this paid-up body of science, I doubt that one could have drawn together the clinical and epidemiological threads which have made it possible to connect an outbreak of Kaposi's sarcoma among homosexuals of New York to infections with *pneumocystis carinii* in Haitians, heterosexual drug abusers, and hemophiliacs.

Unfortunately, Peter Justice and the less publicized victims of AIDS are not only at risk for contagion and cancer, but also for the historic role of the scapegoat. It is by no means to our credit that this affliction of the young and able has rekindled the stale fires of unreason. Nevertheless, I am rather optimistic as to what the future will bring with respect to this epidemic. As a transmittable disease, AIDS looks very much like the sort of contagion the etiology of which we should be able to pin down rapidly. All the early data suggest that it is caused by an infective agent which traverses the blood stream and is transmitted in blood, saliva, or semen. Populations that have experienced AIDS have also seen the virus of hepatitis B or the Epstein-Barr virus. And the tenacious sleuths of the CDC (Center for Disease Control, Atlanta) have shown a consistent pattern of AIDS transmission by means of intimate contacts with known carriers. These lines of information indicate that AIDS is caused by the kind of disease-producing agent we are already quite good at snooping out, and that it is not due to novel forms of social apostasy on the part of its victims. Early students of the disease seem to have been misled by the gay connection: it no longer seems useful to postulate that semen absorbed *per os* or *per rectum* has unique immunosuppressive properties. Besides, it is unlikely that the lymphocytes of hemophiliacs, heterosexual drug abusers, and Haitians have been blemished by heterologous sperm. Finally, I hope that we can soon dispense with the sort of undocumented geographical disputes which seem to be distant replays of the early days of syphilis. When we find the infective agent of AIDS we should be able to staunch the flow of acrimonious letters to *The New England Journal of Medicine* in which the debate has raged as to whether Haitians brought the disease to American gays, or whether gays made the Haitians sick.

If we agree that Secretary Heckler's request is justified, not only for the sake of Peter Justice and his brothers but also on behalf of a common pool of science, let me add another item to the secretary's budget. This summer, while requests for AIDS research were featured on the front pages, another epidemic was documented in back-page snippets among the ads for stereo equipment and savings banks. The cure of this epidemic illness will cost less than the conquest of AIDS; its victims are another minority. Effie Albright, 76, of Woodson Terrace, Missouri, was found dead by the police in her apartment with the windows tightly shut. Joanne Smith, 68, of East St. Louis, was found dead in her home on the same day. These were only two of twenty-nine deaths—among 300 stricken—during the week of July 15–22 in the Southeast and Middle West. Two hundred deaths had been reported by mid-August, and in the last epidemic year (1980) for which the CDC had final tallies, 1,000 had died. The disease is hyperthermia, or what the press calls "heat-related death," and it results when the body's systems for dissipating heat cannot keep up with its generation; these systems function less briskly in the elderly. A spokesman for the CDC in Atlanta, interviewed by *The New York Times,* acknowledged that each year "heat-related deaths" are reported in geographic clusters after continental temperatures exceed 100 degrees Fahrenheit for three to four days; this year, St. Louis and Chicago were particularly affected.

The disease kills mainly the aged and infirm, usually in urban settings. Patients of public nursing homes or homes for the elderly are particularly susceptible; I can recall dog days in the old Bellevue—before air conditioning—when almost everyone over sixty examined on afternoon rounds ran a fever of undetermined origin. It also affects mainly the poor. One afternoon last summer, as temperatures simmered above 100 degrees Fahrenheit, dozens of Atlantans were felled by the heat as they waited patiently in line to receive the free cheese we seem to be handing out these days in place of food stamps.

But this epidemic—we are entitled to use the term if we are to be consistent with our application of it to AIDS—attracted no great demand for public intervention. Indeed, the reports of human deaths received meager newspaper coverage, and were outnumbered by tales of animal disasters and agricultural inconvenience caused by the heat. On July 22, the Associated Press reported that "the worst hot spell in three years has killed a dozen people in Missouri . . . and nine in Georgia. More than 300 cattle perished in South Dakota." A police department statistician explained why the prime victims of the heat were elderly people trapped in brick homes in neighborhoods where they feared to open the windows: "St. Louis has a high crime rate and a dense population, and people might hesitate to open their windows." By August 24, officials in Ohio said that federal agricultural disaster declarations would be sought before the 1983 harvest was complete: their pleas were based on injuries to cattle and crops.

This disease can, of course, be readily prevented; there are groups of individuals who will never die a "heat-related death." Nowhere in the United States last summer was it reported that a bank teller, an airline terminal employee, or a neurosurgeon, for that matter, had succumbed to the heat wave. Folks like these are not usually aged, their places of work are cool, and they are unlikely to "catch" heat exhaustion from their elderly fellow citizens.

Heat deaths in the aged are clearly due to a lack of air conditioning in the presence of poverty. We can do little about age and poverty in the short run, but we ought to be able to buy cool air. Indeed, it seems reasonable to suppose that we can come up with appropriate, prophylactic therapy for those susceptible to heat death by simply providing an air conditioner to those among the elderly too poor to buy their own. Since the only obstacle to this plan is its expense, a little cost accounting may again be in order. Bought in bulk, a perfectly splendid air conditioner should cost $350. Add another $50 for electricity, and the price is $400. An air conditioner should remain trouble-free for at least five years: that brings the yearly cost over half a decade to $80 per person. If two share a room, that's $40. Compare that to the $10,000 of AIDS money per victim!

Now let's determine how many of the at-risk elderly would benefit from government largesse equal to that proposed for the fight on AIDS. Remember that the tab for AIDS research is going to run $40 million per year. Should we be unlucky enough not to have come up with anything definitive for AIDS in five years, that adds up to a cool $200 million for 1984–89. We can air condition a lot of homes and rooming houses for that kind of money: half a million to be exact! That may not be enough to cover the entire susceptible population—I don't have the U.S. Census report at hand—but it would surely make a dent. Recall that the people at risk are the poor, retired, and out-of-work in our South, Southeast and Middle West; we need not provide breezes for their contemporaries in La Jolla, Palm Beach, or Bar Harbor. And if we ask from where the money will come, I'm sure that the back pockets of Secretary Heckler's department (or Secretary Weinberger's, for that matter) might yield more funds from "new furniture and equipment" to our enterprise. If flooding and the death of cattle can open the purse strings of "agricultural disaster" funds, what about the human disasters of Effie Albright and Joanne Smith?

Why is it that we are so ready to spend millions on the conquest of a new disease, but reluctant to do a little good housekeeping on an old one? At least three reasons come to mind, and it is difficult to decide which among them is primary. First, it is more than likely that our attitudes toward AIDS are influenced by a variety of Freudian concerns for our sexual safety and identity. The disease began, after all, in the gay community and it is difficult not to make at least an unconscious association between a nasty contagion and personal or moral "uncleanliness." By throwing money at the problem, we not only wipe our symbolic hands, but hope that the target—AIDS—will go away before it touches us. No such compelling myths shape our attitudes toward the shut-in poor of hot St. Louis: we can't catch anything from them.

Second, we are simply not as good at taking care of each other's commonplace needs as at exploring the new. In the days of plague and typhus, Columbus and Magellan sailed the seas at public expense, while a good part of the population of Europe lived in shacks surrounded by raw sewage. More recently, we have sent bright spaceships to the moon, while in the centers of our cities the poor have frozen in winter and fried in the summer. Driven by natural or economic law, we are happier when we stab at the unknown than when we arrange the fabric of social justice. We have conquered polio but not poverty, tuberculosis but not truancy, syphilis but not slums. Money for AIDS research will flow along the clean and accountable channels of public funding; our social engineers have not fashioned similar conduits through which money can run to cool the poor. Somehow, we seem condemned to triumphs of biological wizardry and failures of social management.

Finally, many of us in medical science have become disinterested in the tedious process of social amelioration. In diverting our interests from the philanthropic aims of our profession, we resemble the Duchesse de Guermantes, described by Proust as

> fearful lest the conversation should turn to philanthropy, which she found boring.
>
> [Said the Duchesse:] "How can one do good to people one doesn't understand? And besides, one doesn't know which people to do good to—one tries to do good to the wrong people. That's what is so frightful."

It is not beyond our capacity, I should have thought, to do good both to Peter Justice and to Effie Albright; they are not the "wrong people." We can make a start by understanding just what it is that is killing them.

Addendum

This article was written in the summer of 1983. Nothing that has happened since would serve to contradict its major points. Unfortunately, AIDS is increasing in incidence: in January of 1985 there were approximately 7,500 cases reported worldwide, with over 3,000 fatalities; no one seems to have recovered from the disease. Fortunately, on the other hand, the prediction made in the summer of 1983 was almost immediately validated, when a group of French researchers reported isolation of the etiologic agent, a virus also isolated at the National Institutes of Health in Bethesda and now called LAV/HTLV-3. The virus has now been characterized by the most refined methods of molecular biology, and antibodies found in infected individuals. By January of 1985, it had been discovered that the immunologic deficit of AIDS results from a unique affinity of the virus for discrete surface molecules present only on a subset of immunocompetent cells (the T4 lymphocytes). Not only has the virus been cloned, but the techniques of genetic engineering are now directed at the production of a vaccine, clinical trials of which are promised for late 1985. Most of the basic work which has permitted this incredibly rapid attack—on a virus that was unknown before 1978— has been funded by monies originally devoted to the war on cancer, not by dollars earmarked for AIDS. Many of the research dollars that were allocated for AIDS research have, as predicted, enlarged our knowledge of basic immunology.

Each week, especially here at Bellevue, the tragic toll of AIDS mounts. But although the disease is still with us, real progress is unquestionably at hand: we are advancing on the scientific front and becoming more responsive to the special plight of the people afflicted. I cannot say the same for similar efforts on behalf of the urban poor.

Auden and the Liposome

Visitors to my office at Bellevue Hospital occasionally identify, with surprise, a photograph of W. H. Auden hung amidst the usual diplomas, family pictures, and group shots of house-staff days. The poet is shown next to Erika Mann; their pairing is a minor document of the troubled thirties. The marriage, in 1935, of Thomas Mann's daughter to Auden, a British subject, conferred a happier nationality upon the stateless refugee from Hitler's Germany. In the picture, Auden wears a rumpled lounge suit; his wide lapels and gaping jacket sleeves are signals of the period. His left hand dangles the perpetual cigarette, the right hand drapes an open jacket to display the braces and skirted trousers of a time when:

> *National Service had not been suggested*
> *O-Level and A were called Certs*
> *Our waistcoats were cut double-breasted*
> *Our flannel trousers like skirts.*
> —from "A Toast"

The poet's hair is neatly trimmed, the face is unlined—no hint appears of those Icelandic crags and furrows that in later years were to transform his features into a relief map of the anxious age. His wife-in-name-only appears to be generating a conventional smile for the photographer. She is dressed in a fashionable windowpane frock, topped by a gay, polka-dotted coat. Her scarf is knotted in chic display, the hat is soft and rakishly tilted. The overall effect is of upper academe. Indeed, the young English master and his wife have been snapped by a photographer from the school newspaper of a progressive private school.

There are two reasons why this picture hangs above my desk: the subject and the photographer. Auden's photo reminds me that what I do for a living—medical research—may begin as fun but has a social bite. And since the snapshot was taken by my sometime collaborator, Alec Bangham, I am reminded of a long Cambridge summer in the sixties when he first taught me to form liposomes and which I remember chiefly as an interlude of pure joy.

Auden himself was persuaded that science, like poetry, is a "gratuitous, not a utile, act, something one does not because one must, but because it is fun." However, it is not this aesthetic approach to the doing of science—the approach of a skirt-trousered amateur—that has engaged me in Auden's poetry and prose. I think, rather, that I respond to that mixture of appreciation and fear of modern science which informs so many of his fabrications. He is at once a lyric enthusiast of our profession—a flatterer of the enterprise—and a necessary critic of our social mischief. His oldest friends—Christopher Isherwood, Cyril Connolly—have called him a schoolboy scientist at heart; Stephen Spender acclaimed him as the diagnostician of our fears.

Son of a physician, familiar with the winners of glittering prizes (his phrase) from the laboratories of Oxbridge and the New World, he paid even the least distinguished of scientists an extravagant compliment that is difficult to forget:

> The true men of action in our time, those who transform the world, are not the politicians and the statesmen, but the scientists. When I find myself in the company of scientists, I feel like a shabby curate who has strayed by mistake into a drawing room full of dukes.

Given the political convictions of his youth, so different from his predecessors—Yeats, Eliot, Pound—this generous appraisal should come as no surprise. Auden and his fellow anti-Fascists of the thirties were convinced that the journals of science contained clues to the fellowship of man. Auden believed that the laws of physics govern servant and master alike, and that it was the job of the poet to instruct both in the language of their common history.

> As biological organisms made of matter, we are subject to the laws of physics and biology: as conscious persons who create our own history we are free to decide what that history shall be. Without science, we should have no notion of equality: without art no notion of liberty.

With these attitudes in tow, Auden devoted much of his energy to warning us of the wretched use to which both poetry and science had been put in our time, in decades during which

> *The night was full of wrong,*
> *Earthquakes and executions,*

And still all over Europe stood the horrible nurses
Itching to boil their children
      —from "Voltaire at Ferney"

Auden was ashamed by the extent to which the children of art and science enlisted in the service of brutality, injustice, and moral squalor. Commissioned as a major at the close of the Second World War, he visited in the course of his work the concentration camps where the methods of science were mocked. He spent long evenings in Bavaria recapitulating

> *The grand apocalyptic dream*
> *In which the persecutors scream*
> *As on the evil Aryan lives*
> *Descends the night of the long knives . . .*
>   —from "New Year's Letter 1939"

In the suburbs of Munich—in Dachau—Professors Pfannenstiel of Marburg, Jarisch of Innsbruck, and Linger of Munich had frozen scores of inmates to death and reported carefully detailed autopsies to "proper" scientific congresses. Here too, Professor Beiglbock of Berlin forced Poles and Jews to drink an excess of seawater: descriptions of the victims' hallucinations and heart failures were exactly recorded in what passed for scientific manuscripts. At the Natzweiler camp, Professor Dr. Eugen Haagen—formerly of the Rockefeller Institute—worked to transmit viral hepatitis from prisoner to prisoner and managed successfully to kill several hundreds with experimental typhus. These examples of scientific disgrace were paralleled in the realm of the arts by the complicities of Heidegger, the gangs of Bayreuth and Oberammergau, by the films of Leni Riefenstahl. After such excesses, Auden became persuaded that our best hope lay in the establishment of limits, limits to the collaboration between intellect and the tyrant, best expressed in his "Ode to Terminus," the Roman God of Limits:

> *In this world our colossal immodesty*
> *has plundered and poisoned, it is possible*
> > *You still might save us, who by now have*
> > *learned this: that scientists, to be truthful,*
> *must remind us to take all they say as a*
> *tall story, that abhorred in the Heav'ns are all*
> > *self-proclaimed poets who, to wow an*
> *audience, utter some resonant lie.*

This plea for limits seems appropriate to our new era of biological engineering and belletristic extravagance. But I sense, perhaps, another strain here: a restatement of "Without science, no equality." For Auden is speaking to us from the experience of a generation which had used the discoveries of physiology and biochemistry as a kind of shield against the biological determinism of the old Fascists. He correctly discerned in the ongoing genetic arguments based upon insect behavior (first ethology, now sociobiology) a nasty trend toward the spinning of tall tales of inequality:

> *Bestiaries are out, now*
> *Research has demonstrated how*
> *They actually behave, they strike us*
> *As being horribly unlike us.*
>
> *Though some believe (some even plan*
> *To do it) that from Urban Man*

By advertising, plus the aid
Of drugs, an insect might be made.

No, Who can learn to love his neighbor
From neuters whose one love is labor
To rid his government of knaves
From commonwealths controlled by slaves?
　　　　　　—from "Bestiaries Are Out"

These verses anticipate my own misgivings about recent attempts to offer the stunning successes of modern biology—our ability to decipher the social code of bees and the genetic code of man—as excuses for undoing the notion of equality. We have detailed the biochemical errors which cause blacks to suffer from sickle cell anemia, Italians and Greeks from thalassemia, Jews from Tay-Sachs disease, and Nordics from pernicious anemia; such heritable flaws speak of biological inequality. Since the popular geneticists of race and behavior have become persuaded that social characteristics also reside in the genes, is it any wonder that neo-Fascists, such as Alan de Benoist of France, have seized upon the recent hypotheses of sociobiology and "selfish genes" to legitimize their political fantasies? Auden's worries *matter:* the tall stories of our most recent science are beginning to have a dirty fallout. Private discoveries have public consequences. That homily brings me to the second reason for the picture on my wall, which evokes not only a summer of fun but yet another worry that Auden did not live to articulate.

As I've said, Auden and Erika Mann were photographed by A. D. Bangham, F. R. S., of Cambridge. In 1935, Alec Bangham was photographer for his school newspaper at the Downs School and had been assigned to photograph the

English master and his new bride. The photo remained imprinted upon a glass negative, stored among the juvenalia of this gifted amateur photographer, until Alec produced it at a scientific conference in the English countryside held fifteen years after the discovery of liposomes. It was at this conference that half a dozen investigators agreed that the use of liposomes in the treatment of human disease was not only desirable, but imminent. Liposomes are small fatty vesicles, made in the laboratory from off-the-shelf chemicals, which Alec and his collaborators originally proposed as models of cell membranes. They were soon found to duplicate many of the properties of the natural bilayers of lipid which enclose the ferments and nucleic acids of living cells. Since liposomes are biodegradable and not at all toxic, it has been suggested that they might function as the long-searched-for vectors by means of which entrapped substances might be safely delivered to organs deep in the body. It took no great effort of the imagination on the part of the conferees at this liposome meeting to tell each other tall stories of the use of liposomes for the manipulation not only of disease, but of the genes. While such experiments are only on the drawing board right now, it may not be premature to put a second concern on the agenda of angst.

I am afraid that in the decade and a half since liposomes were first constructed, developments in biological engineering have come so far, and so fast, that we are on our way not only to explaining, but to perturbing, the fundamental properties of living things. In this decade, when schoolboys and stockbrokers know how to assemble genes in the lab, we have learned to worry not only about the political consequences of our theories, but about the biological *sequelae* of our experiments. We have launched on an endeavor which may eventually realize the prophecy of Diderot:

> If anyone wants to describe . . . the steps in the production of man or animal, he will need to make use of nothing but physical agencies . . . eat, digest, distill in a closed vessel, and you have the whole art of making a man.

Indeed, the distilling of lipids in a closed vessel is a fair description of how liposomes are fashioned. We begin by dissolving lipids, fatty materials identical to those which our own cells use to fabricate their membranes, in chloroform, and then dropping this solution into a closed vessel. The chloroform is evaporated off and the fats remain as a turbid, dry film at the bottom of the vessel. Next comes an operation which approaches magic, and which, each time I perform it, carries with it faint intimations of the Book of Genesis. We add a watery solution which contains any one, or several, of the purified large molecules of life: enzymes, hormones, genes. Then, in obedience to the laws of physics and chemistry, the fats spontaneously enclose these molecules in membranes of predictable geometric array. The suspension assumes an opalescent sheen as the membranes—part liquid, part crystal—swell with their cargo. By simple separative procedures we can then isolate liposome-entrapped materials from those which have escaped capture. Eventually we can hold in our hand—or at least in the collection flask—lipid-entrapped enzymes or nucleic acids: things arranged very much as they would be in a jumbled, rudimentary cell or organelle.

I've described the formation of these little vectors of enzyme or gene in detail because they constitute one example of how the playthings of the lab have suddenly become capable of arousing not only aesthetic joy, but moral qualms as well. Indeed, our colleagues of molecular biology, who can now stitch genes in the dish and harness bacterial energies for the production of human proteins, are now engaged in efforts at introducing their genetic artifacts into the cells of mouse and man. Some of them are probably already toying with the use of liposomes as vectors for bioengineering. I suppose that when I express moral qualms at this possible application of our discovery, I'm only saying that neither Alec nor his collaborators signed up for this sort of activity when liposomes were first made—when one first enthused about the prospect of actually replicating a fundamental unit of life, a membrane capable of entrapping the stuff of cells. Now, there is every reason to believe that in the decades to come, when second- or third-generation liposomes can be appropriately designed to pass safely through body fluids and deliver their contents to vital organs or tumors, their use will prove of benefit in the treatment of disease. But every increment in our capacity for fiddling with the nature of things should, I believe, make us pause for moments of serious self-doubt, should make us worry that we are not only engaged in mischievous tinkering.

The current mistrust of scientific research stems, in my view, from four major insults to our moral sensibilities. I've already alluded to two of these: the unfounded confusion of modern genetics with social Darwinism, and the not unrelated abuse by German doctors of human experimentation. To these may be added two others: the consequences to our offspring of the "poisonings and plunder" of the earth (radioisotopes and chemical pollutants), and now, fears as to the restructuring of man by the well-intentioned splicing of his genes. These accusations are not unfounded, and as a community, we in the sciences should be prepared to acknowledge our share of the guilt, without being hobbled by the admission.

It may be scientific hubris to worry about the small contribution of liposomes to the game of genetic roulette—this will probably be played whether or not our stake is critical to the transport of inheritance. The current prospects for changing our natural load of disease and aging still seem somewhat dim—and we cannot be certain that strategies based on gene splicing or liposomal delivery are even headed in the proper direction. The enterprise is only *about* to be launched. But although the enterprise itself merits concern, I do not believe that we should back off. When I was first in Cambridge, I was present at a fastidious discussion between E. M. Forster and some young transatlantic visitors on his essay entitled "Two Cheers for Democracy." (The mostly radical visitors seem to have been concerned that Forster was giving one cheer too many.) Well, at this point, I'd like to sound two cheers for biology!

It is certainly *possible* that errors and disaster will accompany our attempts to alter the biology of man: but that biology includes diabetes, childhood leukemia, crippling arthritis, and inexorable senescence. Our technical triumphs may change the matter of the natural world, but that world maintains pandemic influenza, endemic parasites of gut and liver, and the natural carcinogens of plant and virus. In the days before the early bioengineers of microbiology (Pasteur, Ehrlich, Koch, and Metchnikoff) began to manipulate the fundamental nature of an ecosystem composed of man and microbes, the natural world contained smallpox, diptheria, poliomyelitis, tuberculosis, and cholera.

A socially prescient humanist of 1880 might well have worried about the future of man in a world freed of microbes by the microbe hunters. He would have warned us of overpopulation, of an aging populace, of consequent famine, inflation, and social unrest. While acknowledging, in partial guilt, that some of these consequences flowed from the discovery of antibiotics, would we have been wiser *not* to conquer infectious diseases? My answer, since I am now alive thanks to antibiotics, has to be negative. Like it or not, on an actuarial basis, you, dear reader, and I are alive thanks to a society that permits the risk of error and invention, that encourages private inquiry to be expressed as public gain: the whole shooting match of Western invention and activism. Those lucky enough to be supported on the playing fields of science should worry hard—before, during, and after the game—but the rules of our sport, of science, are not written in the language of our guilt. The language which we need to remind us of that guilt remains the language of the artist, the poet, the philosopher, and—ultimately—the citizen. Auden has suggested that the language of science and that of poetry are at opposite poles; we neglect the latter at our own peril, because:

> Scientific knowledge is not reciprocal like artistic knowledge: what the scientist knows cannot know him.

So I suppose that I am reassured by that picture on my wall, as I see Auden and Erika Mann looking at the photographer, the young Bangham, and through him at me. As Auden's image in the photograph overlooks my laboratory impedimenta: the many journals, monographs, and reprints, I am persuaded that he knew what scientists are up to, that he knew the extent to which we are guilty, expressing this in language as clear as the genetic code:

> *This passion of our kind*
> *For the process of finding out*
> *Is a fact one can hardly doubt*
> *But I would rejoice in it more*
> *If I knew more clearly what*
> *We wanted the knowledge for,*
> *Felt certain still that the mind*
> *Is free to know or not.*
>
> —from "After Reading
> A Child's Guide to
> Modern Physics"

A Fashion in Metals

COPPER bracelets are sprouting like crocuses in spring. Clasped on the thick wrist of the Seventh Avenue swinger, dangling from the ulna of a tan hairdresser, peeping from the Sea Island cuffs of the broker, their appeal breaches the barriers of age, sex, and political faith. These bangles, which according to folk wisdom can either prevent or ameliorate arthritis, were quite popular among the Aquarians of the sixties and are resurfacing in the let-them-eat-cake reign of Nancy and Ronald.

One is forced, however, to arrive at two rather separate conclusions as to the efficacy of the bracelets, depending on whether one encounters them during off hours or in a medical setting. On the one hand, based on observations of the social scrimmage—the dinner party, the cocktail scene, the promenade—it would appear that folk wisdom is correct: the copper bracelet seems to prevent crippling arthritis. For it is quite obvious in these arenas that the dapper folk who sport copper on their wrists do *not* appear to suffer from arthritis of any sort: their joints appear as mobile as their lives.

On the other hand, in the crowded arthritis clinics of Bellevue, where the indigent gather, afflicted with osteo- and rheumatoid arthritis, with gout or bursitis, copper bracelets are even more frequently encountered than in the salon or the streets. I daresay that close to one-fourth of our patients have tried the device. Indeed, so prevalent is the recourse to manual copper by the arthritic that, were one to cast an hypothesis relating the bracelet to disease, one might well postulate that the wearing of copper bracelets *induces* crippling arthritis.

There is a further paradox in this business. It has been well demonstrated that patients with rheumatoid arthritis have excess copper in their blood—due no doubt to levels higher than normal of a defensive, copper-carrying protein (ceruloplasmin). In fact, at least one approved antirheumatic drug (penicillamine) has the property of leeching copper from the system. Thus, the bracelets, if effective, are unlikely to work by providing, via the skin, traces of the metal lacking in tissues of the arthritic.

In the anecdotal tradition of my clinical subspecialty, I therefore decided to question some of the advocates of the copper panacea, hoping to learn from them how they imagined that the bangles worked. The first chap I encountered, at a soirée, was a fifty-six-year-old businessman wearing a tarnished bracelet of no great beauty. He assured me that for several years his right hand had been so crippled that he found it impossible to decapitate two-minute eggs.

"But after a week or so of this bracelet—it's an antique—I could move my wrist like this," he explained, accompanying his remarks with undulations commonly associated with the Indonesian dance. In truth, the agile wrist fluttering before me was free of any stigmata of arthritis.

"Do you have to wear it on the affected part?" I asked.

"Of course, it's the electric or chemical vibes that free up the joint," he claimed.

A different story was forthcoming from a dealer in primitive art, strolling between the ice-cream parlors on Columbus Avenue. He was very proud of a shiny, elegant bracelet discreetly embossed with the name SABONA.

"If you're really in the know," he confided, "you get these in England. They only cost about two pounds sterling, and, of course, only a SABONA is *O.K.*" This information was accorded with a knowing wink, as if he had just explained to a young Patagonian the meaning of intertwined Gs on a handbag, or the secrets of the initials YSL.

"But do you have arthritis of the wrist?" I asked.

"Oh no, I wear this for a terrible pain I get at times in my right great toe," a digit he waggled at me after extricating it from his topsiders.

"And it works all the way down there?"

"Of course: copper bracelets make you feel good all over. Anyone knows that they cure arthritis!"

Keeping my own association with the arthritis field strictly under wraps, I questioned a score of others, and was repaid with glowing accounts of the therapeutic utility of copper bracelets for tennis elbow and gout, for bunions and osteoarthritis, for housemaid's knee and rheumatoid arthritis. Some were persuaded of their specific local effects, others of more diffuse powers. A poor woman with systemic lupus erythematosus told me that it "charged her whole system" and helped to regrow her depilated hair; a prisoner with gonorrhea suggested that the bracelet not only tamed his swollen joints but "made him more of a man again"— and so on.

Another strain of credulity became evident. One young lady told me that her bracelet worked only when Jupiter was in the ascendency; a serious journalist assured me that since he was a Leo, the bracelet was most effective in the summer. Indeed, a strong association was appreciated by many between the healing power of the metal and the conjunction of the stars as revealed in the truths of astrology. From somewhere in the collective unconscious of these acolytes of copper emerged a vigorous belief that there was a clear connection between the planets in orbit and their metal talismans. On closer questioning, this took the form of a clear faith that the electrical forces generated by heavenly bodies were transmitted to their joints by way of the copper bracelet.

These quaint notions might well be dismissed as the aberrations of a few. But it was not too long ago that an American president—the "nucular engineer," as he pronounced the word—claimed not only to have had an epiphany in the woods, but also to have observed a UFO, and that the astrologer Linda Goodman was welcomed to the White House by our Chief Executive. There are, we are told, 20,000 professional astrologers in these United States— as opposed to 2,000 astronomers. Belief in the karma cycle unites the young ecologists of the ashrams with the devout of Orange County. Meditation unites the professional linebacker with the social activist. Chiropractors and faith healers are welcome practitioners on Capitol Hill. Indeed, in this antirational climate, it is difficult to determine whether an amendment to repeal the Enlightenment would not fare better than the ERA.

The strong association between the power of copper bracelets and the energy of the planets reminded me of another story of metals and arthritis, one with a more rational conclusion, perhaps. The story is that of "saturnine gout." Gout is caused by the irritative property of crystals of monosodium urate in tissues of the joint. Consequent to elevations of the level of soluble uric acid in the plasma of those affected, crystals of monosodium urate precipitate in the joints and cause acute, painful arthritis, not infrequently of the great toe. In most cases of gout, there is a strong genetic basis for overproduction or undersecretion of uric acid, which is an end product of purine metabolism in man. However, in cases of saturnine gout, which is due to chronic lead poisoning, uric acid accumulates because the metal blocks uric acid excretion by the kidney.

Two "epidemics" of saturnine gout are recorded in medical history. The first of these afflicted chiefly the propertied classes of England in the eighteenth and early nineteenth centuries. The second, to this day, affects poor blacks in rural areas of the southeastern United States. Both are due to the accidental contamination of alcoholic beverages by lead: fortified wines were the vector for lead in Britain, moonshine is laced with metal in Alabama. In such diverse places and in two diverse populations, the sign of Saturn heralded the agony of gout. Why is this condition called "saturnine" gout? We come back to links between astrology and the joint. In alchemical parlance *Saturn* is the name for lead. To medieval astronomers, Saturn was the remotest of the planets, revolving most slowly in its sullen orbit. From these observations, the astrologers derived the notion that those born under its sign were cold, sluggish, and baleful of temperament. Lead—that heavy, sluggish metal—became *Saturn* by extension.

By the eighteenth century, the adjective "saturnine" was applied to those of ruddy complexion, of heavy weight, of substantial social position, of a tendency to take the gout. From the early eighteenth to the mid-nineteenth century, the letters and graphic art of England documented a close relationship between alcoholism and gout among the gentry. Yet for reasons unknown in the eighteenth century, neither the social peers of the English upper classes—the drunken lords of Germany, Sweden, France, and Russia—nor the lower classes of England seemed to suffer from a similar prevalence of the gout. The answer lay in observations made later in the nineteenth century. Clinicians gradually became aware that it was fortified wines such as port or Madeira which seemed to bring about the gout. Indeed, A. B. Garrod—the "father of gout"—attempted in 1859 to identify chemical differences between wines that provoked attacks and those that did not. Lacking the analytical tools, he was nevertheless able to establish that light wines or whiskey were relatively harmless, but that fortified wines from the Iberian peninsula or the Canaries were conducive to gout. He drew no connection between this observation and his other finding that fully one-quarter of his gouty patients were exposed to toxic doses of lead.

Recent studies by Gene Ball of Birmingham, Alabama, seem finally to have placed these observations of Garrod into a narrative frame. Reasoning that the lower classes of England drank gin, and that the continental gentry quaffed mainly unfortified wines or vodka, he analyzed the oldest available bottles of port and Madeira for their content of lead. Ball drew on his extensive knowledge of arthritis and his equally extensive experience of fine wine. With the aid of Michael Broadbent, the expert at Christie's on rare vintages, he came in possession of four rare bottles of port, Malaga, and Canary Island wine. Dating from 1770 to 1810, they had been sequestered in the cellars of a Scottish duke. When the wines were analyzed for their content of lead, it was found that they contained as much as 1,900 micrograms per liter—as compared to modern vintages of similar provenance which contained no more than 180 micrograms. Their lead content, he reported, by no means interfered, in respect to taste, with their "clear superiority over the less venerable products." Ball postulated that the metal was derived from lead-reinforced tubing which was used by the Portuguese during the distillation of brandy, which, in turn, is the added ingredient of fortified wine. He was supported in his distillation hypothesis by the predominant cause of gout among the black population of rural Alabama—"moonshine gout." In the southeastern United States, illegal distillers frequently utilize old automobile radiators as condensers. The resultant brew, free from federal taxes and retail mark-ups, is heavily enriched in lead solubilized from soldered radiators. Chronic refreshment by this homemade liquor causes the drinkers to suffer from slow lead poisoning—and from moonshine gout.

An excursion into the history of English drinking practices thus led Ball to an explanation for the gouty epidemic of the eighteenth and early nineteenth centuries. In 1703, the English Foreign Minister, Methuen, had signed a favorable trade treaty which made it possible to import fortified wines from the Portuguese mainland and from the

Canary Islands at two-thirds the price of French wines. In due course, large "tuns" (252 old wine-gallons) of port and Madeira, rich and flavorful despite the long sea voyage, began to arrive in the Home Counties. Indeed, the brandy which fortified Portuguese wine not only provided greater flavor, but, in part because of the additional ethanol, rendered it capable of transport to northern ports without spoilage. Shortly thereafter (1723) came the first report by Musgrove of "saturnine gout." In the event, consumption of fortified wines by well-off Britons became so great that the diet of noble Albion was appropriately described as consisting entirely of beef and port. By 1825, the English were importing 40,277 tuns of port per annum. The results of this century-long debauch are pictured in song and story, and in the memorable engravings of Hogarth and Rowlandson.

But it was the Prince Regent, later George IV, no stranger to port or gout, who made the disease—and swilling—a *sine qua non* of fashionable living. His first attack, in 1811, followed a long boycott of the products of French vineyards. His subsequent tussles with the disease were liberally and affectionately documented in broadsheets and popular engravings, as Eric Bywaters has lavishly illustrated. George IV seems, almost by personal example, to have capped the eighteenth century's common equation of gout and abuse of fortified wine with exalted social rank. When the English spoke of "the Honour of the Gout," the touch of gout became a touch of class.

Saturnine gout became a bit more diffused among the lower English social strata as the century progressed. A new source of dietary lead became prominent. By 1879, the reports of St. George's Hospital described "two patients who took 1½ pints of beer per meal (both saturnic)" and "Three patients who took 2 pints of beer (one was gouty and saturnic)." In these cases the beverage was unwittingly flavored by lead which had escaped from pewter mugs. Indeed, saturnine gout was recognized as an occupational hazard of "Potmen, who drink beer that has stood for some time in pewter vessels." So the disease of the exalted soon became an affliction of the lower classes, who ingested with cheap beer that which was once reserved for their social betters. Yet the mechanism for the induction of gout remained a mystery until the disease made a strong comeback in Alabama.

Ball and James T. Halla, writing in "Seminars in Arthritis and Rheumatism," discuss the magnitude of the problem at the Birmingham Alabama VA Hospital. The drinking of moonshine accounted for seventy-nine cases of lead intoxication. Of these, forty-two had gout, and the forty-two patients constituted 36 percent of the total cases of gout seen in the hospital from 1970 to 1976. Ball and Halla report that analyses of samples of moonshine made in Alabama during the sixties showed toxic levels of lead in 40 percent of the brews. Thanks, however, to a variety of factors, including improved law enforcement and public education, moonshine is rapidly declining in favor and the authors happily report that, at present, saturnine gout accounts for only 5 percent of the cases of gout in Birmingham. The disease is making a slight comeback in another part of the country, however. James Klinenberg of UCLA tells me that orange juice, due to its acidity, leaches lead from ceramic jugs brought by illegal immigrants from Mexico. Saturnine gout has flowed from George IV to our own Chicanos. How democratic the disease has become!

But in this continuing story, the images that keep recurring are those of the person of George IV. Visibly disabled by the gout, saturnine, fat, and elegant, he appeals to us—or appalls us—as a kind of sacrifice on the altar of fashion. Felled by the leaden rot of his kidney, he succeeded nevertheless in making his disease synonymous with the good life: a punishment, perhaps; a distinction, certainly. Indeed, I have a hunch that were saturnine George to appear on our television screens today—full of port, staunch of temperament, riddled by gout—he could persuade us of almost anything. The mixture of danger and rank, disease and fashion would certainly grab us today: it was only a few decades ago that we all donned trench coats and chainsmoked cigarettes in the fashion of Bogart or Edward R. Murrow.

When I think of good George, I am also brought in mind of our own trivial gentry: the fashionable Biancas, Marisas, Geoffreys, and Oscars who lounge about the pages of *Vogue* prattling astrology. For there, among glossy photos, in manors of unreason, from under taffeta, silk, crêpe de Chine—there where the diamonds and ivory dangle—one can observe, oh so often, the glint of the copper bracelet. In an engraving of the 1820s, George is depicted on his throne: he is holding audience. Surrounded by courtiers, confronted by supplicants, he is flanked by wigmaker and tailor. But the devils of dropsy and gout nibble at him; he looks like a wigged Robert Morley selling the Honor of Gout at reduced air fare. How silly he looks! But he becomes positively imperial in aspect when one places this engraving next to a new *Vogue,* from the pages of which a suntanned designer of expensive jeans—a Sagittarius—flashes his copper bracelet.

The Chart of the Novel

LIKE other professionals—football scouts, diplomats, and underwriters come to mind—doctors write many words, under pressure of time and for a limited audience. I refer, of course, to the medical charts of our patients. Most of these manuscripts (for even now this material is written almost entirely by hand) are rarely consulted by anyone other than doctors, nurses, or (Heaven forbid!) lawyers. I have always found the libraries of this literature, the record rooms of hospitals, to contain a repository of human, as well as clinical, observations. On their tacky shelves, bound in buff cardboard and sometimes only partially decipherable, are stories of pluck and disaster, muddle and death. If well compressed and described, these *Brief Lives* are more chiseled than Aubrey's. Best of all, I love the conventional paragraph by which the story is introduced. Our tales invariably begin with the chief complaint that brought the patient to the hospital. Consider this sampling from three local hospitals:

This is the third MSH admission of a chronically wasted 64-year-old, 98-pound male Hungarian refugee composer, admitted from an unheated residential hotel with the chief complaint of progressive weakness (1945).

This is the second SVH admission of a 39-year-old, obese Welsh male poet, admitted in acute coma after vomiting blood (1953).

This is the first BH admission of a 22-year-old black, female activist transferred from the Women's House of Detention with chief complaints of sharp abdominal pains and an acutely inflamed knee (1968).

How evocative of time, place, and person—and how different in tone and feeling from other sorts of opening lines! They certainly owe little to the news story: "Secretary of State Schultz appeared today before the Foreign Relations Committee of the Senate to urge ratification of . . ." Nor does the description of a chief complaint owe its punch to any derivation from formal scientific prose: "Although the metabolism of arachidonic acid has been less studied in neutrophils than . . ."

The opening sequence of a medical record is unique, and when well written, there's nothing quite like it. These lines localize a human being of defined sex, age, race, occupation, and physical appearance to a moment of extreme crisis: He or she has been "admitted." Attention must therefore be paid, and everything recorded on that chart after the admitting note is a narrative account of that attention, of medical "care."

But this enthusiasm for the products of clinical prose may be unwarranted. There may be other prose forms which, by their nature, tug at the reader with such firm hands: I have not found them. Now, my search has not been exhaustive—indeed, my inquiries are based on a sort of hunt-and-peck excursion amongst the yellowing survivors of my dated library. I must report, however, that no similar jabs of evocative prose hit me from the opening lines of biographer, of critic, of historian—not at all the kind of impact I was looking for. No, the real revelation came from the nineteenth-century novelist. I should not have been surprised:

> Emma Woodhouse, handsome, clever and rich, with a comfortable home and happy disposition, seemed to unite some of the best blessings of existence; and had lived nearly twenty-one years in the world with very little to distress or vex her.
>
> —*Emma*
> Jane Austen

> Madame Vauquer (nee Deconflans) is an elderly person who for the past forty years has kept a lodging house in the Rue Neuve-Sainte-Genevieve, in the district that lies between the Latin Quarter and the Faubourg Saint-Marcel. Her house receives men and women and no word has ever been breathed against her respectable establishment.
>
> —*Le Père Goriot*
> Honoré de Balzac

> Fyodor Pavlovitch Karamazov, a landowner well known in our district in his own day, and still remembered among us owing to his mysterious and tragic death, was a strange type, despicable and vicious, and at the same time absurd.
>
> —*The Brothers Karamazov*
> Fyodor Dostoyevsky

Now, that sort of beginning is a little more like the opening paragraphs of our charts. Does this mean that we've been writing nineteenth-century novels all our professional lives, but without knowing it? Do the white, pink, and blue sheets which describe the events between admission and discharge—between beginning and end—constitute a multi-authored *roman à clef?* If we go on to the rest of the record, it is more likely than not that other sources can be identified. The "History of Present Illness," with its chronological listings of coughs, grippes, and disability, owes much to the novelist, but more to the diarist. But the "Past Medical History," drawn to the broader scale of social interactions, returns us again to the world of the novel, and the more detailed "Family and Social History," which describes the ailments of aunts and of nephews, which lists not only military service but the patient's choice of addiction, puts us into the very middle of the realistic novel of 1860.

Now comes the "Physical Examination." Here the world of the clinic or the laboratory intrudes: numbers, descriptions, and measurements. Indeed, from this point on, the record is written in recognition of the debt medicine owes to formal scientific exposition. In this portion of the chart, after all the histories are taken, after the chest has been thumped and the spleen has been fingered, after the white cells have been counted and the potassium surveyed, the doctor can be seen to abandon the position of recorder and to assume that of the natural scientist. He arrives at a "Tentative Diagnosis," a hypothesis, so to speak, to be tested, as time, laboratory procedures, or responses to treatment confirm or deny the initial impression. The revisions of this hypothesis, together with accounts of how doctors and patient learn more about what is *really* wrong, constitute the bulk of our manuscript.

So we can argue that our records are an amalgam between the observational norms of the nineteenth-century novelist and the causal descriptions of the physiologist. There is, perhaps, a connection between the two. If we agree that a novelist not only tells a story but weaves a plot, we imply by this the concept of causality. E. M. Forster, in *Aspects of the Novel,* draws the distinction nicely. He suggests that when a writer tells us, "The King dies, and then the Queen died," we are being told a story. When, however, the sentence is altered to read, "The King dies, and therefore the Queen died of grief," we are offered a plot—the notion of causality has been introduced. In much of our medical record keeping we are busy spinning a series of clinical plots: the temperature went down *because* antibiotics were given, etc.

Pick up a chart at random, and you will see what I mean. The sixty-year-old taxi driver has been treated for eight days with antibiotics for his pneumonia. "Intern's Note: Fever down, sputum clearing, will obtain follow-up X-ray." A few days later: "X-ray shows round nodule near segment with resolving infiltrate. Have obtained permission for bronchoscopy and biopsy." Then, "JAR Note: Results of biopsy discussed with patient and family." Months intervene, and at the end of the readmission chart we find the dreary "Intern's Note: 4:00 A.M. Called to see patient . . ." Infection and tumor, hypothesis and test, beginning and end. And so we read these mixtures of story and plot, learning as much along the way about the sensibilities of the doctor as we do of the patient and his disease. The physician-narrator becomes as important to the tale as the unseen Balzac lurking in the boarding house of Madame Vauquer.

An optimistic attempt to reconcile both sources of our clinical narratives, the novel and the scientist's notebook, was made by that great naturalist—and optimist—Emile Zola. After an exhilarating dip into the work of Claude Bernard, Zola decided that the new modes of scientific description and their causal analyses might yield a *method* which would apply to the novel as well. Basing his argument on Bernard's *An Introduction to the Study of Experimental Medicine* (1865), Zola wrote an essay entitled *The Experimental Novel.* Zola explained that the novelist customarily begins with an experimental fact: He has observed—so to speak— the behavior of a fictional protagonist. Then, using the inference of character as a sort of hypothesis, the novelist invents a series of lifelike situations which test, as it were, whether the observations of behavior are concordant with the inference. The unfolding of the narrative, interpreted causally as plot, will then naturally, and inevitably, verify the hypothesis. How neat—and how reductionist!

But however simplified this scheme of Zola may appear to us today, it has the merit of suggesting how strong, indeed, is the base in scientific optimism upon which traditional clinical description rests. Our descriptions imply our confidence that detailed observation of individual responses to common disease have a permanent value, which can be used predictively. They reveal an upbeat conviction that causal relations, when appreciated, lead to therapeutic (or narrative) success. Recent views of medicine and the novel seem to challenge these assumptions.

If we look at the ways in which we have changed our

records in the last decade to the "Problem-Oriented Patient Record," to the "Defined Data Base," we appear to have shifted from a view of patient and disease based on the human, novelistic approach of the last century to one based on the flow sheet of the electronic engineer or the punch card of the computer. No longer do our early narratives end in a tentative diagnosis, a testable hypothesis: we are left with a series of unconnected "problems." The stories dissolve into a sort of diagnostic litany—e.g., anemia, weight loss, fever, skin spots—without the unifying plot that ties these up with the causal thread of leukemia. Worse yet, these records are now frequently transformed into a series of checks scrawled over preprinted sheets which carry in tedious detail a computer-generated laundry list of signs and symptoms. The anomie of impersonal, corporate personnel forms has crept into these records. Added to these forces, which have turned the doctor's prose into institutional slang, is the movement to eliminate reflections on sex preference, race, and social background. In the name of convenience and egalitarianism, we seem to have exchanged the story of the single sick human at a moment of crisis for an impersonal checklist which describes a "case" with "problems." When we fail at words, we fail to understand, we fail to feel.

But, I'm afraid that the new novelists have anticipated us here, too. As the naturalistic novel has yielded to the stream of consciousness, to existential angst, and to flat introspection, the anomie of the clinic has been foreshadowed by that of the artist. The opening lines of our major modern novels sound the tones of disengagement as clearly as our clinical records:

Today, mother is dead. Or perhaps yesterday. I don't know. I received a telegram from the Home. "Mother dead. Funeral tomorrow. Best Wishes." It means nothing. Perhaps it was yesterday.

—*The Stranger*
Albert Camus (1942)

If only I could explain to you how changed I am since those days! Changed yet still the same, but now I can view my old preoccupation with a calm eye.

—*The Benefactor*
Susan Sontag (1963)

What makes Iago evil? Some people ask. I never ask.

—*Play It As It Lays*
Joan Didion (1970)

Perhaps as doctors we are now committed to acting as a group of "benefactors" ministering to the sick "strangers"— we cannot, or will not, be involved in the lives of those who have come to us for care; we will now simply describe and solve the problems of the case. We will play it as it lays.

The Urchins of Summer

For nearly a hundred summers, biologists have gathered near the deep channel that joins Vineyard Sound to Buzzards Bay at Woods Hole. Most of these scholars have at one time or other performed the classic experiment of the Marine Biological Laboratory: They have fertilized urchin eggs with urchin sperm in a small dish. Sea urchins, or, to give them their formal name in Massachusetts, *Arbacia punctulata,* are ubiquitous in these waters; their fertile season happily overlaps the long vacation of academic calendars. Unlike their counterparts in other climates, *Arbacia* yield no great pleasure to the palate; their utility to man seems limited to their role as a substrate for summer curiosity. The wet, bristling sea urchin is a creature of no conventional beauty; like the porcupine, its surface repels our touch. Protected by countless spines, encased in umber shells, embroidered with delicate tentacles, the prickly *Arbacia* resemble ripe walnuts with protruding quills.

But it is not the adult sea urchins that have brought us

all to the seashore. We come to study their gametes. The female urchins, plucked from gurgling sea tables in the laboratory, shed clusters of claret-colored eggs into beakers after a bit of harmless salt water is injected into their body cavities. And under the lowest power of a light microscope, it is possible to follow the stirrings of life in the garnet interiors of the ova. When a drop of sperm is added to a suspension of eggs, scores of the barely visible, motile fellows surround each ovum. They scuttle about in brisk competition; only one will connect with any one egg.

Once sperm and egg have joined, the developmental clock is set: within two minutes, refractile halos appear around each egg. These shimmering fertilization membranes prevent the egg from penetration by other sperm. By forty or fifty minutes the eggs, in reasonable synchrony, furrow, and then cleave—where each membrane enclosed one egg, one cell, we now observe two cells. Soon after, the cells divide again—first to form a Maltese cross of four, then eight, sixteen, and so on. The cells become uncountable and their clear geometry disappears. By the next morning, the dish is filled with motile urchin embryos, called *plutei,* which swim about in random celebration of their right to urchin life.

Over the years, seaside biologists have studied this glittering ritual with all the cunning of their modern craft. From urchins, we have learned how ions govern the life of cells, how sperm mingle their genes with those of the egg, how DNA makes RNA, how genes are amplified, and how the cytoskeleton braces the machinery of organelles. Our journals are filled with intricate accounts of the budding *Arbacia* of Woods Hole.

Persuaded that the activation of urchin eggs holds lessons for the cell biology of inflammation, I have spent over a dozen seasons watching this junction of gametes. Indeed, my summers can be described as that period of the year when I pay equal attention to tiny spheres of *Arbacia* and to larger, yellow spheres labeled WILSON 4. In June of 1982, while waiting for the urchin eggs under my microscope to progress from fertilization to first cleavage, this attention was repaid by yet another sort of lesson. Temporarily unoccupied, I turned to the morning *New York Times* to discover the news that a committee of the U.S. Senate was seriously engaged in elaborating a definition of the beginning of human life; the junction of sperm with egg.

What a coincidence! On my left, urchin life was stirring in the dish. On my right, the newspaper of record informed me that our government had entered into the legislation of biology. I had always assumed that human, as opposed to *Arbacia,* life began somewhat later. Indeed, I have tended to agree with Talmudic doctrine, which teaches that the human fetus becomes a viable creature only when it graduates from medical school. Further on in the newspaper, I discovered that the committee's deliberations had been aided by a number of eminent physicians and advocates from the Right to Life movement. Most of them argued that human life began when sperm joined egg and that such a position was based on "incontestable" biological fact. It turned out that both Lewis Thomas, presently at Stony Brook, and Leon Rosenberg, of Yale, expressed some misgivings as to whether any such thing as the beginning of life could be subject to a "scientific" definition. These scholars proposed that such a definition had best be left to ethicists, philosophers, or theologians. I gathered that the issue was charged with emotional and political content—this was no mere academic skirmish. For it appeared that most of the adherents of the "sperm-meets-egg" definition were outspoken proponents of legislation designed to prohibit abortion on demand or at least abortions sought by the poor.

There were still thirty or so minutes to go before the urchin eggs were due to cleave. I began to worry. I reckoned that each summer I must have studied anywhere from ten to fifty thousand urchin eggs, all of which were carelessly flushed down the drain somewhere in the course of transit to the free-swimming stage. Was I guilty of running a marine abortion mill? If urchin life—like human life—begins at fertilization, have I been guilty of mass murder? Would pickets from the Animal Defense League in Falmouth suddenly appear on Water Street, joined by brigades of "cat-ladies" dragged from their stations at the American Museum of Natural History? How far down the great chain of being would congressional fiat extend? Would the Senate be lobbied into defining the beginning of cat life, of urchin life—and what about *E. coli?* Well, not being a professional ethicist, philosopher, or theologian, I had only personal experience to draw on for consolation. And I suddenly remembered having faced a somewhat similar set of questions.

In the summer of 1960, I had arrived at the Strangeways Research Laboratory in Cambridge, England, to begin a postdoctoral fellowship. Since I had proposed a study of how papain and vitamin A affected the structure of cartilage, I inquired of the director of the laboratory, Dame Honor Fell, where one might obtain the rabbits needed for such experiments. Our group at NYU, led by Lewis Thomas, had found that the ear cartilage of rabbits underwent dissolution when the animals were fed an excess of vitamin A or were injected with papain, an enzyme best known as a meat tenderizer.

In Cambridge, however, there was a problem. It appeared that one needed a special license in order to work with animals. Ordinarily, this was readily granted to bona fide scientists, but the local university administrator had died suddenly. Months might elapse before his yet-to-be-appointed successor would be available to approve my animal adventures. I went to the appropriate Home Office representative and asked if, unlicensed, I could work on bacteria? Yes. Could I experiment with amoebae or paramecia? Yes. Fish? Of course! Could I work with amphibians? Yes, but only before metamorphosis—after this transformation they were classed as *real* animals and came under The Act!

That was my first brush with the legislative approach to the morality of phylogeny. For someone in the further reaches of the Home Office—someone with the best intentions in the world—had decided that the protection of animals extended to frogs and toads, but not their precursors. In the event, I found that the tadpole larvae of South African clawed toads *(Xenopus laevis)* also suffered dissolution of cartilage when exposed to an excess of vitamin A. The year was not lost; by the time a license came through I was too busy feeding nettles to my tadpoles to bother with rabbits.

Recollecting the experience twenty years later by the waters of Massachusetts, I decided that the Home Office had gotten it almost right, after all. There is a difference between urchin or toad life and the life of mammals, a difference that depends upon another "incontrovertible" fact of biology: Urchins or toads drop their eggs in the water, mammals nurture them in the womb. The fertilized egg of the rabbit or human requires the life of its mother for sustenance until, at birth, the individual is ready for an independent life of its own.

Not for naught have the pro-choice forces chosen the rusty coat hanger as their symbol. I recalled the many frightened, confused young women brought bleeding into the emergency room of Bellevue Hospital before abortions were legalized. Victims of amateur or semiprofessional instrumentation in the tenements or barrios of the city, many of these injured girls—the majority were abused teenagers—had become septic or were near the point of shock. Unlike their wealthier sisters, who could afford the services of more savory, if illegal, practitioners, some of the women lost their wombs in the process. In the most awful sense, any future children they might have borne were robbed of the right to life. It was this vision of a gynecological abattoir, in the era before legalized abortions, that came to mind from the pages of the *Times*.

I turned back to the urchins under the microscope. Division was in full swing: the eggs had cleaved and the little spheres shimmered in their gossamer shells. Soon the eggs would be on their way to urchinhood. In the saline world of *Arbacia*, union of sperm and egg was not only necessary but *sufficient* for the viability of the young. Is this true of human life as well? Now, I've already disclaimed expertise in professional ethics, but the problem of necessity versus sufficiency seems a real one. One can argue, I suppose, that human life-of-a-sort begins when a fertilized egg is implanted in the mother's womb. But this *necessary* event is not *sufficient* for human life. Given the present-day medical technology, most physicians would assert that an embryo requires at least the better part of the gestation period before independent life is possible. Within that time span, the two lives are one, the smaller dependent entirely on the well-being of the larger.

At the end of the nineteenth century, doctors and midwives would have agreed that babies born before the final month of pregnancy would have precious little chance of surviving *ex utero*. As our science has progressed, we have moved the period of possible survival further back toward the moment of implantation. And, I daresay, given luck, skill, and social approval, we may get to the point where the required period of uterine nurture may become infinitesimal. But that is to beg the question. Most women will bear most babies for nine months. And in the course of gestation two independent lives are admixed; the mother's womb guarantees *sufficiency* of the child's life. Unlike the sea urchin, the human embryo is not a free swimmer in an open ocean. The onset of urchin life is probably correctly defined as the moment when sperm meets egg. For the human, this act is *necessary* but not *sufficient*. For this reason, at least, I cannot help but believe that there is no way to ground a definition of human life in facts common to human and marine biology. If human life differs from that of the sea urchin, its definition must surely include the facts of urchin gametology, but not be limited by those facts. Biologist or doctor, ethicist or legislator—each, I should have thought, is entitled to work out such a definition according to his social or religious predilection.

The arguments for or against abortion will be with us for some time; the abuse of these arguments by the zealots of the right wing is, one can hope, only a temporary swing of the political pendulum. Sitting at the seaside, I found no great lesson for this social controversy in the joining of urchin eggs with urchin sperm (a process more readily quantified than its human counterpart). The "facts" of biology teach us only about the natural world, not about the moral order. I can readily sympathize with those who place the sanctity of any human life—even that of the fertilized egg in the womb—at the top of their code of moral values. But I remember too many of the victims of rusty coat hangers, of furtive abortionists, or of tumbles down the tenement stairs—their fractured wombs and broken lives—of the era before legal abortion. The urchins of summer remind me that there are two lives involved in the bearing of human children; the facts of biology do not tell me which life to cherish over the other. My social urges and a liberal conscience persuade me to support a woman's freedom of choice.

Poussin and the Bomb:
Metaphor as Illness

I HAVE been worrying about the bomb, and the worries started in the most unlikely of settings. In late August of each year, the long afternoons on the beaches of Cape Cod are lit by a clear, rose light, which transforms the figures of this shimmering landscape into the classic tableaux of Poussin. On one of those crystal afternoons, as the beach on Vineyard Sound was almost emptied, a group of the long-haired young—students and laboratory assistants at Woods Hole—had dispersed themselves around a large ice chest. The total arrangement suggested languor and peace. Or so it seemed to me, until, in passing by, I deciphered the peeling bumper sticker plastered on the chest: NO NUKES.

Here then, in this Arcadia of academe, was a reminder of the great unspoken threat to our time—a *memento mori,* so to speak. "Even here in Arcadia . . ."—a familiar phrase, but not so familiar that I could place it with immediate certainty. So, sand still between my toes, I plodded back to our summer home, where the sun-bleached paperbacks

of yesterday yielded at last the great essay by Erwin Panofsky on "Poussin and the Elegiac Tradition." There I found
an evocation of the whole scene on a faded page. Panofsky
displays the two paintings of Poussin that bear the title
"*Et in Arcadia Ego*": an early version now at Chatsworth
and a later one in the Louvre. In the first, painted around
1630, two curlyheaded young shepherds accompanied by a
ravishing shepherdess have stumbled upon a chestlike sar-
cophagus, propped in the midst of Arcadia. The legend
decipherable on *this* chest is "*Et in Arcadia Ego.*" Panofsky
calls our attention to a small death's-head, a skull, perched
on the tomb. In a dazzling display of scholarship, Panofsky
unravels the iconography of Poussin by tracing the theme
of Arcadia to Virgil, the grammar of the inscription to its
only permissible meaning ("Even in Arcadia, there am I
[death]"), and the skull to its antecedents in medieval ver-
sions of the *memento mori*. But the bulk of Panofsky's essay
is devoted to contrasting this early version (which was evoked
for me on Nobska Beach) with one in the Louvre painted
less than a decade after the first. The shepherds in the later
version are

> symmetrically arranged on either side of a sepulchral mon-
> ument. Instead of being checked in their progress by an
> unexpected and terrifying phenomenon, they are absorbed
> in calm discussion and pensive contemplation. One of the
> shepherds kneels on the ground as though rereading the
> inscription for himself. The second seems to discuss it with
> a lovely girl . . . and the death's-head is eliminated . . .
> The Arcadians are not so much warned of an implacable
> future as they are immersed in mellow meditation on a
> beautiful past.

I find this passage a suitable description of our own shifting attitudes to the *memento mori* of our times, the skull on our tomb, the atomic bomb. Our earliest reaction to the warnings of an implacable future was the straightforward cry of "Ban the Bomb!" But we seem to have mellowed as the stockpiles have grown, and the best among us have adopted the more comfortable slogan, "No Nukes." We have given terror a nickname. Many of us, like the young of Woods Hole (and the Arcadians of Poussin), have stopped worrying about the death's-head and learned to love, or at least live, with it. The worn banners of "Ban the Bomb" have been stashed; bumper stickers and T-shirts demand "No Nukes" mainly to protest the use of nuclear reactors for local energy. We have transformed outraged response into a meditation on ecology, a threat to Arcadia.

Is there a more direct resonance between our increasingly relaxed view of nuclear weapons and the tranquil, elegiac tradition of Poussin? Panofsky suggests that the tranquilization of the *memento mori* theme is "consistent with the more relaxed and less fearful spirit of a period that had triumphantly emerged from the spasms of the Counter-Reformation."

Indeed, Poussin grew up in that period of relative religious tolerance ushered in by the Edict of Nantes (1598). Poussin seems to have been captured by the Arcadian theme during his studies in Rome, where the Mannerist Guercino first painted an "*Et in Arcadia Ego*" in 1621. In this Mannerist painting, we are shown less of Arcadia and more of death, its composition dominated by a mottled skull. Guercino's painting is more closely related to the traditional "reminder of death" motif than is Poussin's. The *me-*

mento mori found its first, and grimmest, expressions during the "Waning of the Middle Ages," as Huizinga has reminded us. The later medieval world responded to pandemic plague (in 1348) by elaborating images of death, frequently represented as a dancing skeleton amidst the young, the strong, and the fair. With time and the Renaissance, the reminders became less insistent. The next flurry of necrography coincided with the bitter doctrinal wars of the sixteenth century, but in the Mannerist style of the Counter-Reformation, death had already been reduced to the single skull.

By the age of Poussin, the wars of religion had largely subsided, and the "less fearful spirit" of the time permitted Poussin to move the metaphor of death from skull to tomb (death was losing its sting). Poussin's two treatments, therefore, not only bridge the stylistic gap between Mannerism and his version of the baroque but also introduce us to the world of the seventeenth century—the world of metaphor. Roger Fry has drawn analogies between the concerns of Poussin and those of John Milton. The dialectic, in Milton, between flesh ("L'Allegro") and spirit ("Il Penseroso") appears in Poussin in the form of encounters between lyric sherpherds and Grecian tombs, metaphors remote from the more direct—and more horrid—images of the dance of death.

In the language of our age, are the metaphors for death and the bomb more direct? During the early stages of our ongoing religious conflict—the Cold War—the possibility of imminent nuclear war seemed very real to the people of my generation. The awful titles of the time (*On Nuclear Warfare. Thinking the Unthinkable*) reflected that fear, and there was real debate as to whether survival was possible or worthwhile. The words were tough, and thanks to Bertrand Russell and Linus Pauling, the "Ban the Bomb" movement created a climate for the test-ban treaty. In contrast, our recent campaigns for a nuclear "freeze" seem, to me, occupational therapy for the Volvo set. We should be in the streets, and scared to death! But now, in the dull prose of our warrior-scholars, we read of such diluted euphemisms as "clean bombs," and "first-strike capabilities," of "strategic arms limitation," and of the deployment of such acronyms as MIRVs and MXs. One plan for playing with MXs was presented by Carter's people as a kind of tinker-toy, defensive trolley. Those leaders, who seemed to have trouble running functional commuter trains, told us—believe it or not—that the security of the Free World was to have been served by a set of nuclear missiles that rolled on tracks around rural silos. The plans of Reagan's generals for the MX—his Buck Rogers fantasies for outer space—are cheerful movie effects properly described as "Star Wars."

While the generals are engaged in verbally disguising the terrors of nuclear three-card monte, expressions of angst, or dread, in our serious culture have also received a cosmetic treatment. As Poussin replaced the image of skeleton or skull with that of the manicured tomb, so have the artists of our day reduced the threat of atomic extinction to more acceptable (and diluted) metaphors.

The fear of nuclear warfare and its expression in metaphoric terms was the subject of a slightly boozy conversation at one of our occasional poker games. My friend, the psychiatrist, dealing me two low pairs, told us of his clinical experience with the "No Nukes" generation: "They're really worried about the bomb. The ecology movement and the fear of nuclear reactors are surrogates for the real terror—a terror they've lived with all their lives."

This clinical observation had been made before. The Yale psychiatrist Robert Jay Lifton has pointed out that for many of his patients the battleground of the unconscious is no longer the family romance. The Freudian repression of sexual guilt has been replaced by the angst of our time (of nuclear annihilation)—dreams of terror and explosion. The results of this shift had worried the late Arthur Koestler:

> The explosions produced a kind of psycho-active fallout which works unconsciously and indirectly, creating such bizarre phenomena as flower-people, drop-outs and barefoot crusaders without a cross. They seem to be products of a mental radiation sickness, which causes . . . an existential vacuum, a search for the place of value in a world of facts. But in a world that refuses to face the facts there is no such place.

And after my wife raised the betting on the basis of a hidden straight, she suggested some of the metaphors for destruction that have been produced in popular and serious culture: the overt themes of crippling, deformity, and helplessness in such popular dramas as *The Elephant Man, Whose Life Is It Anyway?,* and *Wings*—not to speak of the latent, serious terrors of Beckett *(Happy Days, Endgame).* Once on this tack, it was not difficult for the other players to suggest that metaphoric use of the nuclear *memento mori* was a feature of much postwar art. We identified mushroom clouds in Morris Louis' "Veils," disasters in the black scaffolds of Franz Kline, and white explosions in the "POW" and "CRACK" of Lichtenstein's pop art. Baselitz, Schnabel, Cucchi, and Kieffer—the new expressionists—seem to be painting the landscapes of a nuclear *Totentantz.* Recent forays into the American novel of fact have given us *In Cold Blood* and *The Executioner's Song*—the presence of death in the absence of affect. Nor did we fail to throw into this cultural hopper the popularity of horror and disaster films: tales of earthquakes, volcanic eruptions, Australian floods, and American sharks. Fashion was not exempt: the torn shards of Issey Myake's post-Hiroshima sweatshirts and the ragged skirts featured by *Comme des Garçons* seem appropriate clothing for the disaffected young. Indeed, it was not long before every cultural display of deformity (from the abattoirs of Francis Bacon to the beasts of Borges) became another example of a metaphor for terror.

Well, this uncritical sort of free association that frequently accompanies a game of cards may or may not bear rigorous examination. After all, themes of mutilation and images of horror in the arts certainly antedate Hiroshima. But I cannot help feeling that what has been left out of our cultural production is important. We have not really faced the awful music of our time. The pop nightmares of Jonathan Schell, the reassuring face of Jason Robards on "The Day After" are essentially trivial responses. Nuclear terror has not been given a persuasive local habitation and a name. To paraphrase Marianne Moore, we have placed imaginary toads, instead of real ones, in the gardens of our popular and serious arts. If this be so, we should not be surprised that the efforts to "Ban the Bomb" have faded into a "veggie" crusade to close Three Mile Island.

In our kind of world, the slogan "No Nukes" is a fuzzy, misdirected one. Even the hawks of Southern California love this nickname, "Nukes." It was in La Jolla that I saw the ultimate spoof of the movement on another bumper sticker: NUKE THE GAY WHALES. No, if we cannot distinguish between atomic energy and its terrible use in warfare, we are in deep trouble. As I write this in Bellevue Hospital, the isotopes of iodine, technetium, hydrogen, and carbon are tracing the patterns of disease in patients down the hall. The sick are treated in ways and by means that would have been impossible without the triumphs of cell and molecular biology made possible by the reactors of Upton, Oak Ridge, and Argonne. And to hark back to Woods Hole, it is quite likely that the same young scientists who had gathered about the "No Nukes" chest on the beach had just finished popping their little vials into scintillation counters back in the lab. There is, indeed, in some places, at some times, a coherent and moral argument to be made in favor of nuclear energy for civilian use. Nuclear energy may be an alternative to the prospect of acid rain, for example. No, the slogan of "No Nukes" does not attract me; it seems to be a symptom of an illness that the atomic age has rendered pandemic. The disease is that of evasion—the failure to name, to describe in the clearest way possible, in ways devoid of metaphor, the fact that we may all blow each other up in ways Koestler has suggested:

> These symptoms too, will wear off. But there is no getting away from the fact that from now onward our species lives on borrowed time. It carries a time-bomb fastened round its neck. We shall have to listen to the sound of its ticking now louder, now softer, now louder again, for decades and centuries to come, until it either blows up or we succeed in de-fusing it.

We have come a long way since Poussin, but there remains yet one more bizarre relationship between the painter and the bomb. Another historian has written a fine essay on Poussin. In it, he argues that

> the difference between the two versions of the Et in Arcadia Ego theme lies above all in the fact that in the first there is only regret and disillusionment shown at the transitoriness of life, whereas in the second there is resignation . . .
>
> It was not till the end of his life that [Poussin] came back to the themes which had attracted him in his early years, and then, as we have seen, it was to treat them in a very different spirit, with a resigned detachment, not with emotional disillusionment . . .

This was published in *Art Bulletin* by Anthony Blunt, a distinguished scholar recently unmasked as a fellow conspirator of Burgess and Maclean. Blunt and his fellow spies played a significant role not only in the Cold War but in the spread of the secrets of atomic warfare. One wonders whether he viewed these weapons with "resigned detachment" or with "emotional disillusionment." I suspect that the reports to his Soviet masters were free of metaphor.

So count me out of the "No Nukes" campaign, the lines forming around Indian Point. Our race will not be extinguished, although it may be randomly assaulted, by accidents of civilian carelessness. No, I am persuaded that the real danger to our species is that the wars of religion (East against West) will explode again—not by error (although that is possible), but by intent. This threat is the only one that counts; its abolition is the task of each.

One and a Half Cultures:
The Uses of Medical History

Doctors live in a world of eponyms. We are anxious to diagnose *Addison's* disease, to worry whether a microorganism fulfills *Koch's* postulates, eager to determine whether a child suffers from *Lesch-Nyhan* syndrome, or troubled whether a murmur of the heart is of the *Austin Flint* type. In clinic, laboratory, and at the bedside, we carry with us the labeled baggage of medical history. Most of us are ambivalent about these impedimenta. On the one hand, it is more convenient to remember *Reiter's* syndrome than the clinical triad of tribulations that this eponym describes. On the other hand, we are vexed with our memory because it usually cannot locate the eponymist in place or time. (Who was Reiter? Is it Addison's disease or his anemia?) Minds filled with *new* bits of clinical and scientific data, we rarely have time to examine the historical basis of our language or craft. Unfortunately, it seems that the urge to expose our medical roots becomes insistent only as we progress to professional senescence; in midcareer we are frequently reduced to the quip of S. J. Perelman, who insisted, "I have Bright's disease, and he has mine!"

This state of affairs has been aggravated by the notions current in professional education that the historical approach is irrelevant to technical métiers, and that gentlemanly incursions into the social sciences can produce, at best, only a sort of amateur gossip. Such antihistorical attitudes may be related to two dominant collegiate trends of the 1960s. Students expressed in physical terms their impatience with what were perceived as the elitist biases of liberal education. In complementary fashion, faculties of the arts and social sciences, whilst seeking to raise their own standards of scholarship, have limited the formerly broad appeal of their subjects to the narrow boundaries of professional humanism. Both trends have culminated in the attrition of liberal education in our colleges, in our premedical requirements, and in examinations for entrance to medical school.

Deprecatory views of the value of general education for the young scientist or doctor are not new. Throughout the nineteenth century, medical science was engaged in a kind of revolt against the dry, school-book learning of overly historical scholars who dominated much of European medical education. One thinks of Turgenev, whose character Bazarov in *Fathers and Sons* exemplifies the liberating aspect of this attitude. We forget how revolutionary a gesture it was for Alfred North Whitehead to claim (in 1916): "A science that hesitates to forget its founders is lost."

Ironically, Whitehead was echoing the founders of our own science who accused science in general, and scientific medical education in particular, of paying excess attention to authorities of the past, while showing too little interest in the experimental present. The antihistorical views of scientific education, which became prominent after the First World War, must be appreciated as a kind of revolt against the dogmatic reverence of classical or historical learning. Again it was Whitehead who sounded the tocsin for utilitarians by suggesting that:

> The understanding which we want is an understanding of an insistent present. The only use of a knowledge of the past is to equip us for the present. No more deadly harm can be done to young minds than by depreciation of the present. The present contains all there is. It is holy ground; for it is the past and it is the future. The communion of saints is a great and inspiring assemblage, but it has only one possible hall of meeting, and that is the present; and the mere lapse of time through which any particular group of saints must travel to reach that meeting place makes very little difference.

Distantly sanctioned by such authority, we no longer crowd the medical curriculum with considerations of medical history. Tacitly, we have accepted the argument that a knowledge of the history of his craft is largely irrelevant to the young scientist or physician. Now that we understand, for example, that chronic kidney failure can follow not only infections by the streptococcus but can be induced by chronic pyelonephritis, hypertension, and the renal disease of systemic lupus erythematosus, is it really so illuminating to read the description by Bright of the disease which bears his name? Surely he described a grab-bag of these diagnoses.

But perhaps we should not read the history of medicine or science as a record of rights or wrongs, a series of diagnoses correctly or incorrectly arrived at. Perhaps the utility of past for the present is the lessons it proves in *modes* of analysis, in patterns of recognition, in the way that perceptions alter definition. If we follow the historical development of any one problem in medicine, such as Bright's disease, we will be forced to examine our own theoretical constructs (what we now call paradigms), our own patterns of description, our own definitions. As we vicariously participate in Bright's discovery and its elaboration, we look at our own clinical problems in a new way: we can, if lucky, sense the *possibility* of a new generalization to be plucked from the messy strands of fact. This sense of possibility, or wonder, can probably never be so effectively evoked as by way of our historical sense.

But the usual medical or scientific histories confront us with a paradox. Those sufficiently informed to appreciate the technical basis of renal physiology and pathophysiology (for example) are usually not sufficiently versed in general history to understand the social, cultural, or economic factors which led to the discovery of *this* disease in *that* hospital at *that* time. On the other hand, social scientists equipped with this sort of knowledge are ill prepared to understand the ultrastructural differences between diffuse and lumpy deposits on the basement membranes of the kidney. The rather unsatisfactory state of our history of medicine and science may be due to this sort of "uncertainty principle." Those still active in the sciences are ill equipped to analyze their historical development; those who appreciate historical development are usually incapable of presenting sufficient technical detail to arouse the interest of the trained scientist or clinician.

Does this mean that the history of science or medicine is condemned to the level of biographical anecdote or gossip? Is it peculiar to their role in *science* that Koch was an unhappy general practitioner, that Pasteur suffered from early hemiplegia, that Metchnikoff became obsessed with old age, or that Astley Cooper was guilty of nepotism? Perhaps not, but the biographical approach has the value of engaging the amateur, and it is only as amateurs that most of us can approach our own history. If it is impossible for us, despite a considerable period of university education, to maintain a general level of cultivation in the arts and sciences, we can at least hone these rusting talents on the whetstone of medical history. An aroused sense of the history of our own craft should also assure us of those benefits that have been vouchsafed students of other sorts of history: a deeper understanding of the social context in which experiment, discovery, and their consolidation into the body of knowledge have taken place.

It is likely that, in the course of our quest, we will encounter not only people but social constellations that can inform the present. Thus, for example, an account of the golden era of Guy's Hospital might introduce us not only to Addison, Gull, and Hodgkin, but might also permit us to appreciate why the clinical vision of the English empiricists was able to flourish in the Era of Reform. For it was an earlier shift of the rural poor to crowded, urban warrens of the South Bank (remember the enclosure movement, the Corn Laws?) that permitted assembly of individuals displaying similar signs and symptoms: syndromes. Analysis of such syndromes permitted the clinicians of Guy's to discern in *months* the generality of clinical presentations that heretofore had only been encountered once or twice in *years* of practice in country or at court. The urban teaching hospital also gave rise to the clinical vision in France and led to the *Birth of the Clinic* described by Michel Foucault. One might, in fact, come to appreciate how the great break-throughs of microbiology of the last part of the nineteenth century were made possible by French and German advances in scientific instrumentation and synthetic chemistry. These advances permitted the microbe hunters to utilize chemical and optical techniques in order to isolate such previously unseen vectors of human disease as those which cause tuberculosis, diptheria, and anthrax. Understanding, in the first instance, of the social milieu which gave birth to the clinic and, in the second instance, of the history of technology which led to the triumphs of Metchnikoff, Pasteur, and Koch should permit us to frame questions pertinent to our new era of molecular biology and molecular disease.

Medical history, as viewed from the 1980s, is often presented as a series of individual victories of researchers and courageous rebels over the opposition of erroneous authority. Crudely, we might say that the usual history of science is exclusively that of successful scientists. However, none of these brave discoveries arose *in vacuo*. If we read medical history carefully, those of us incapable of entering the pantheon of discovery can recognize that the past achievements of our craft have been made possible by our quotidian peers of earlier generations: those thousands of physicians and scientists who contributed to the proper setting, providing the oyster, so to speak, for the pearls of success.

A final note: We have been told that we are, in one sense, living in the world of the "two cultures"—that of the arts and that of the sciences. The realm of the arts has been suffused by the perception that for the last twenty or so years its creative spirits have been dissipating themselves in a kind of sterile obsession with the self. The process of self-discovery and self-analysis, so the argument goes, has pretty much obliterated the capacity of the arts to engage real people in an external world. This process, which Lionel Trilling has associated with a narcissistic drive for "authenticity" seems, indeed, to have made it possible for us these days to speak of "one culture and a half." Since it is unlikely that we can ever bridge the gap between our scientific culture and that of the other half, perhaps we should avoid a similar dilution and attrition of our side. One threat to our still intact scientific culture has been alluded to before: The drive toward antihistory and worship of the present, energized by the notion that the only viable contributions to science and medicine are those which appear in the last issue of *The New England Journal of Medicine* or *The Journal of Molecular Biology.* This view, to my mind, is the scientific equivalent of excessive self-involvement in the arts and humanities. One can hope that we are not in danger of succumbing to this trend. And this hope is based on the unproved belief that medical science, of all the sciences, is kept honest by our daily encounters with real people who are afflicted with real diseases. When we bring to our practice the baggage of eponymic medicine, we carry with us historical precedent, modified by our own clinical experience and shaped by the science of the present. Indeed, as practitioners of this historical craft we can be said, in Trilling's words, to be:

> Creatures of time, we are creatures of the historical sense not only as men have always been, but in a new way. Possibly this may be for the worse; we would perhaps be stronger if we believed that Now contained all things, and that we in our barbarian moment were all that had ever

been. Without a sense of the past we might be more certain, less weighted down and apprehensive. We might also be less generous, and certainly we would be less aware. In any case, we have a sense of the past and must live with it, and by it.

Golems and Chimeras

RESOLVED that there are some things biological scientists ought not to know, because if they know them, our sense of what is human will be violated." That, in its simplest form, is the subject of ongoing debate not only among biologists but in the U.S. Senate, symposia on ethics, and in the columns of periodicals that range from *Reader's Digest* to *The New York Review of Books*.

Until recently biomedical scientists have been quite certain that any new knowledge they acquired about the nature of living things could not help but be useful to the general welfare, and that such "utilitarian" values constituted a moral guarantee for their intrusions upon the natural world. In consequence, they only had to follow their "aesthetic" concerns in pursuit of the elegant experiment, the beautiful proof, the unshakable theory. Recent events have unsettled this agreeable conviction.

There is ample support for the idea that "science-for-its-own-sake," the aesthetic view of science, yields utilitarian benefits. Research looking not so much for a cure for disease, but rather seeking the nature of soil fungi, gave us streptomycin. Inquiry into the nature of cells in culture led to the Salk vaccine. And studies of the cell cycle in onion root tips have suggested a rational treatment for leukemia. Such examples, and at least threescore others which decorate our recent history, have reassured biological scientists that the enterprise is intrinsically benign. As a corollary, scientists have persuaded themselves that if they perform their task professionally, by doing well, they do good.

His doubt relieved as to ends, the biomedical scientist has concentrated on means. And what magnificent means they are! The toy shops of technology, and the purses of our government, have provided the biologist with electron microscopes which show single cells peeled and split like oranges, centrifuges which hurl viruses at gravitational forces a hundred thousandfold that of the Earth, X-rays which display the molecular symmetry of our tendons, and spectrophotometers which scan the uncoiling of our genes.

By these means, the game of science has been played in obedience to a set of rules that has remained uncluttered by any ethical stricture save one: *Thou shalt not fudge the data.* The professional code of the scientist has been a stringently aesthetic one. It has rewarded the individual imagination for coming up with reproducible experiments. And the scientific imagination has, until recently, been considered by one and all to resemble that of the creative artist, no more—and no less—in the service of transient, social mores.

W. H. Auden summed it up this way: "Both science and art are primarily spiritual activities, whatever practical applications may be derived from their results."

But the biological revolution has now become so strongly challenged that the aesthetic values of biomedical research have been edged into disrepute. Indeed, even the utilitarian ends of science—the manipulation of nature for the eradication of what we perceive as its errors—have been attacked by environmentalists, humanists, and the new theologians. Energized a decade ago by the folly of our technology in Vietnam (organic herbicides against the tropical forest, psychoactive drugs in the hands of the CIA), critics from without have been joined by the disenchanted young from within the perimeter of science, culminating in the latest fuss over recombinant DNA. What began as a brave internal effort to face the ethical problems raised by gene splicing has slowly developed into a broad social movement to proscribe certain kinds of inquiry. For the first time in recent Western history, there is a good chance that some, perhaps benign, authority will legally declare to the biological scientist: Thou shalt not do this experiment, because it is *morally* wrong to muck about with our genes.

Rather than rehearse the scenarios of gloom that opponents of DNA recombinancy have plotted, let us use the categories of Harvard biologist Bernard Davis, who claims that they retell three popular myths: the Andromeda-Strain fantasy, the legend of chimeras, and the creation of the Golem. Nor will I go into the overwhelming evidence that these myths are unlikely to be realized. Indeed, they become more improbable with each new issue of *Science* or *Nature,* as the solid experimental advances of June render pointless the January complaints of legislator or lobbyist.

Instead, let us turn to a friendly critic, June Goodfield, in whose book, *Playing God,* is recounted the history of the genetic debate. Goodfield, elaborating upon its larger meaning, admits that ". . . it is so hard to produce a *rational* argument for one's moral qualms about DNA research." She nevertheless encapsulates the less-than-rational ones which are so much more powerful:

> What bothered us so about the new technology? Three things came to mind: the slow erosion of that which up to this point in our history has gone to make us uniquely human, or what we have considered to be human; the latter-day assault on personal autonomy and integrity; and the increasing sense that individuals are losing control over the conduct and direction of human affairs.

I hope it is not fanciful to point out that the first of these worries is a restatement of the Golem fantasy, the second a rephrasing of the chimera myth (assault on personal integrity), and the third reflects a modest confusion between *Brave New World* and *The Double Helix.* But this reduction of Goodfield's arguments does not render them less cogent. Scientists share with humanists of her persuasion these worries as to the remote, unpleasant, utilitarian consequences of gene therapy, even as they exult over their recent capacity to turn bacteria into engines for the production of insulin or somatostatin.

But the most striking extension of such social and ethical anxieties is yet to come, for as the philosopher A. J. Ayer has suggested: ". . . ethical terms do not serve only to *express* feelings. They are calculated also to *arouse* feeling, and so to stimulate action [my emphasis]."

The action which is called for seems to me to conflict with both the utilitarian and aesthetic values of biomedical research. For Goodfield and others suggest that the time has arrived, perhaps, to impose, by means of external authority, a *professional* code upon the scientist. Such a code, arrived at by the usual democratic process, and with the usual degree of consent by the regulated, will guarantee— we may presume—that after their successful pursuit of the Golden Fleece researchers shall not return home to Medea. These professional codes (as fairly arrived at as the codicils of the Internal Revenue?) will not only enlarge the number of degree bearers now legislatively responsible for the products of their endeavors, but also submit these folks to moral scrutiny.

Among the concerns of legislators and ethical philosophers one can pick out the clear message that the moral imperative for regulation comes from fear of technological abuse. And the chief abuse of which biologists stand accused is their capacity to fiddle with genes. The reformers go on to urge not only that professional scientists should be governed by some sort of professional guild—like the American Medical Association or the American Bar Association—but that they should be subject to the laws of malpractice.

Goodfield describes the unregulated state of science before our recent, ethical concerns:

> Save for the expenditure of society's funds, however, the [scientific] profession was still accountable for nothing. The law in no way held them to the highest degree of care: they were never sued for malpractice nor for misapplication of their work. The only set of ethical principles that ever concerned them were those concerned with protecting the good name of the profession and its "sublime" methodology. They were in no way concerned with the needs of society . . .

This argument assumes that the utilitarian output of scientists has some unique implications for the social order, and that these implications require society at large to set some limit on their aesthetically motivated quest. But are the social consequences of scientific research different in kind from those of other scholarly pursuits or from the creative arts?

Professional historians, for example, suggest that the utility of their narratives provides us with an understanding of the present; professional philosophers often believe that the usefulness of their métier leads us to proper conduct; and social scientists agree that their analyses are of use in the amelioration of political problems. But can we say with certainty that the contributions of Fichte, Nietzsche, and Spengler had less grave consequences for the orderly flow of social progress that even the most lurid results imaginable of our biologic inquiries? Should the aesthetic motivations of Pound, Céline, or the Marquis de Sade have been modulated at the source by a morals committee? Cyril Connolly has quipped: *"Le coeur a ses raisons*—and so have rheumatism and the flu."

Since some of us look into the reasons of rheumatism for the reasons of our heart, is it extravagant for the sciences to claim—at least with respect to subject matter—a measure of the license granted to the arts?

Like it or not, we are discussing censorship when we liken the profession of science to the profession of law or of medicine. Malpractice rules are written to ensure that the practitioner conforms to the general level of professional practice in the community—surely this is *not* the standard of the creative scientists, nor of the innovative scholar in any field. The possible misuses, in the utilitarian sense, of his knowledge should not, I should have thought, provoke preemptive rules as to the kind of inquiry in which the professional engages. Scientists, in the best interpretation, are not *only* professionals in the sense that lawyers or doctors are. In the best sense, they are professionals in the way of historians, poets, or artists.

Let me advance this claim, as phrased by Auden: "Liberty is prior to virtue, i.e., liberty cannot be distinguished from license, for freedom of choice is neither good nor bad, but the human prerequisite without which virtue and vice have no meaning. Virtue is, of course, preferable to vice but to choose vice is preferable to having virtue chosen for one."

Such a completely libertarian view is clearly counter to the temper of our time. We must, indeed, accept the rational bases for the general disenchantment of our culture with the products of science pursued for its own ends. The wonderful folks who gave you (however indirectly) the bomb of Hiroshima, the laboratory of Auschwitz, and the psychiatry of the Gulag archipelago are not generally trusted to keep their new genetic tools locked safely in the academic cupboard.

But from where did the counterforces arise to these overt excesses? To a large extent, scientists themselves, acting in response to a code of moral values they share with others, have blown the whistle on more of these outrages. The atomic scientists, by means of their *Bulletin,* have resisted the proliferation of their monster; biomedical scientists have propagated the Helsinki principles; and the psychiatrists have exposed the social mischief of the mental health commissars. The motivating force behind these ameliorative actions has been, I believe, the sense of humane ethics with which our society at large is impregnated.

Our only guarantee that new knowledge will not be gained at the expense of human values is the integrity of that network of values in which all our inquiries are enmeshed. This is to argue that we should impose no preemptive restrictions on the kinds of new knowledge to be sought. If we wish, for example, to prohibit certain investigations on human subjects by our physiologists because their Nazi counterparts violated any reasonable code of behavior, we may be granting the Nazis a posthumous victory they do not deserve. Of course, science should not design experiments that are trivial, dangerous, or dehumanizing. However, these adjectives are best defined by the consensus of society at large. If that consensus be justly derived, we can expect that the community of scientists, as citizens, will agree.

George Steiner has recently argued that "truth at any price!—unrestricted inquiry into anything at all at any time—is an over-riding cultural value of our sort of civilization." It may well be, but only to the extent that this value does not entirely conflict with those values of society that scientists themselves have introjected. There are many "truths" and many kinds of knowledge—and if we want to be sure that scientists (following their own aesthetic bents) do not come up with chimeras or Golems, we'd be well advised to make certain that our society as a whole doesn't want such fantasies realized. The alternative—the proscription of those lines of inquiry which could possibly lead to chimeras or Golems—is based on the assumption that only the *worst* consequences of an experiment are to be expected. Many of us over forty would not be alive today if that assumption had been dominant in the last century.

Finally, many of us are convinced that the culture of science, which engages at least as much of our population as does the humanist culture, is worth nurturing for its own sake. The unraveling of the genetic code, the elucidation of cell structure and metabolism, the analysis of how nerves make muscles twitch or speech possible, constitute cultural achievements no less imposing than the mosaics of Ravenna or the cathedrals of France. At a time when the arts of our decade are devoted to the production and analysis of works related to self, the objective triumphs of our science are perhaps even more to be cherished. We are, probably, unlikely to achieve the utilitarian ends of our science: freedom from hunger and disease, less anxious youth and sturdy age, without adopting the risky libertarian view that there shall be no bounds as to what scientists need to know. Motivated by aesthetic considerations to reach the aims of utility, new biomedical knowledge will come—in the words of Lewis Thomas, ". . . if the air is right . . . in its own season, like pure honey."

No Ideas But in Things

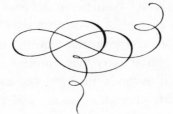

Febrauary is a bad month for doctors in and around New York. Colds hit the young, pneumonia collects the old. Exposure saps the drunk and the homeless. Hospital beds are full and telephone lines are busy; it's dark when we awake and darker still when we get home. It's tough to get around the streets when they're clogged with snow, the roads become caked with brown slush. But no matter how bad it may seem nowadays, it must have been worse a generation ago before the new vaccines and antibiotics. Patients were less willing—or able—to come to the office or clinic and doctors spent more time making house calls, a good bit of that time served behind grimy windshields. If you want a whiff of that sort of winter—of that sort of life—try William Carlos Williams of Rutherford, New Jersey. Robert Coles, the eloquent Harvard psychiatrist, has just brought Williams' medical fictions back into print. But don't stop there. Try his *Autobiography,* or the epic five books of *Paterson.* If you're hooked by then you'll want to revisit his many other volumes of poetry, his novels, essays—you may, in fact, become as preoccupied with this physician/poet as I have been lately. Which brings me back to winter, the winter of 1948, as described by his biographer, Paul Mariani.

The "meadows" of New Jersey, ruined flatlands and industrial suburbs which lie between the Passaic and Hudson rivers, had been covered by snow since the blizzards of early December. At one-thirty on the morning of February 10, Dr. Williams, almost sixty-five, climbed into his Buick, which was parked in the hospital lot. Lot and roadside were banked by yard-high walls of snow, and as he rolled toward home, he playfully took a swipe at the snowbanks with his fenders. The car became stuck in the bank and no amount of rocking would free it. No help was available, so he trudged back to the hospital and borrowed the only tool available, a coal shovel with a broken handle. Setting to work at his usual brisk pace and short temper, he became so furious in the process that he tore his shin badly by kicking the iced snowbank in anger. Hard, mindless work. Williams had stayed late at the hospital with his patients—he'd been codirector of pediatrics since 1931—and coped throughout the day with the chaff of hospital politics. He was president of the Board of Directors of the hospital and representative to it of the Medical Board.

As the doctor shoveled, there was much on his mind. In the past few months, his two careers had come to a turning point. He would soon have to retire from his hospital position. And now that his son, William Eric Williams, was about to finish his pediatric residency at New York Hospital before taking over the bulk of his father's practice, it looked as if William Carlos would finally get a chance to devote more time to his writing. For years, Williams had placed himself on a tough treadmill, indeed. Up early, he'd soon made morning house calls—by no means limited to children—in the ethnic slums which bordered suburban Rutherford. On to the hospital, where he worked till noon. Home for lunch with his wife, Floss: a small nap, then afternoon office hours. More house calls, a delivery or two, dinner, and evening office hours. The sign on his house read:

Office Hours 1 to 2—7 to 8:30
Sundays by Appointment

He wrote poems and prose in the evenings and between patients; sometimes he pulled his car to the curb and jotted down pieces in the little notebook he kept as school physician. He made almost no money at his writing, had his works printed in small editions—which he sometimes supported by income from his practice—and had not yet found a major publisher. Although by 1948 he had achieved a good bit of success among the avant-garde, he was dismissed by most influential critics. Edmund Wilson ranked him with Maxwell Bodenheim as an inconsequential figure and Williams was disappointed that his "friend" Conrad Aiken left him out of a major anthology of American poets. Perhaps now that his son was about to pitch in he could begin to write that long, important work which would gain him the reputation he felt he deserved.

Not that Williams was unknown in the world of arts and letters. Far from it: among his earliest friends were the painters Charles Demuth, Charles Sheeler, and Marsden Hartley. It is no accident that Williams came of age with modern American painting. He maintained a steady correspondence with Marianne Moore, Wallace Stevens, and Kenneth Burke. In Paris he had been photographed by Man Ray, dined with Brancusi, and had circumcised Hemingway's son. He had edited a small magazine with Nathanael West. Ford Madox Ford had founded the Friends of William Carlos Williams Society. Marcel Duchamp had played on the lawn in Rutherford. And within the past few months, Williams had gone over the proofs of *Paterson 2* with the young Robert Lowell and worried with him over the next two sections of that book. He had visited his lifelong friend and political opposite, Ezra Pound, at St. Elizabeth's Hospital in Washington, D.C., where Pound had been incarcerated for madness rather than tried for treason. But his bohemian friends in the arts knew very little of that other life across the Hudson, the life of the solid, hard-working practitioner, who drained abscesses, scraped tonsils, and adjusted baby formulas at two dollars a visit.

Eventually, by this sixty-fifth year of his life, the strain was beginning to tell. He had written to his friend Fred Miller:

> I don't quite understand why I feel so pressed, there seems to be, on the surface, nothing more than I have always handled somehow in the past, more or less successfully, but these days I'm going about in circles . . .

As he continued to shovel, he felt the kind of severe anterior chest pain that any doctor recognizes as either angina or infarction. He rested against the car, experiencing, perhaps in his own body, the angst of Old Doc Rivers, hero of his best short story:

> Frightened, under stress, the heart beats faster, the blood is driven to the extremities of the nerves, floods the centers

of action and a man feels in a flame . . . That awful fever of work which we feel especially in the United States—he had it. A trembling in the arms and thighs, a tightness of the neck and in the head above the eyes—fast breath, vague pains in the muscles and in the feet.

Despite the continued pain, Williams kept digging. Finally, the car began to budge and he was able to get into it. Perhaps only another doctor can imagine what went on that night. Surely, Williams must have known that every textbook of clinical medicine contained a description of the classic candidate for angina or infarction: a sixty-five-year-old male who shovels his car out of the snow in February. And since the pain did not leave him as he drove back to Rutherford he must have been aware that he had suffered a heart attack. But home he drove, not back to the hospital! Home to Florence Herman Williams, his wife since 1912, home to 9 Ridge Road, where his mother Elena had ruled a world of spirits in her invalid room for many of her hundred years.

We have no record of what Williams thought that night, as he returned to home and bed. Certainly the pains persisted as the Buick rolled downhill to Rutherford. Did he expect to die, there in his car, in that mobile world in which he spent so much time and which featured in so many of his stories? The tale of his alter ego, Doc Rivers, begins with: "Horses . . . For a physician everything depended upon horses. They were a factor determining his life." Williams' own multihorsepowered Buick, now carrying his life, had seen these roads before at all seasons and all hours. Its journeys to "Guinea Hill" permitted Williams "entrance to the secret gardens of self"—those encounters to which he attributed the strength of his writing.

147

Many of *The Doctor Stories* begin at the end of a journey by an unnamed physician/narrator to the house of a patient, where the encounter between doctor and patient teaches the doctor—and the reader—something entirely unexpected. Doc Rivers makes many such journeys, takes up cocaine as *his* second career, becomes a small-town deity, and surprises us by realizing the fantasy of every doctor who has ever been worried by house calls: ". . . he built a fine house with a large garden, lawns and a double garage, where he kept two cars always ready for service." In "The Girl with a Pimply Face," the narrator over the course of several home visits to a sick baby becomes aroused by her sister, a nymphet, ". . . Legs bared to the hip. A powerful little animal." The baby is discovered to have a congenital lesion of the heart, the doctor's own heart is engaged by the sister, "a tough little nut finding her own way in the world." Finally, he is astonished to learn from colleagues at the hospital that the girl has a "dozen wise guys on her trail every night." In "A Night in June," the doctor drives across the tracks to the house of an Italian mother of eight. The difficult delivery is helped not so much by his "science"—an extract of pituitrin—but by the comforting, and expert, hands of the patient. He discovers that:

> The woman in her present condition would have seemed repulsive to me ten years ago—now, poor soul, I see her to be as clean as a cow that calves . . . It was I who was being comforted and soothed.

In "Danse Pseudomacabre," our doctor is roused from his bed at three o'clock in the morning by a fat Scot with an infection of the face. The narrator learns that he is needed not only to minister to the patient, but to witness the will! In "The Use of Force," the doctor encounters an attractive young "little heifer" of a thing who puts up a tremendous struggle to prevent her throat from being examined. The girl provokes a rage in the narrator, who is quite aware of the sexual overtones as he confesses how "I could have torn the child apart in my own fury and enjoyed it. It was a pleasure to attack her." In the course of making the diagnosis of the patient's disease—diptheria—the doctor has acknowledged his own.

But the doctor who drove in pain through the Jersey night was not only a writer of conventional short stories in the realistic mode, he was a poet for whom invention and imagination—the discovery of the new—was a variety of religious experience. Perhaps behind the wheel of the car that night, this secular faith was of use. From *Paterson 5:*

> *We shall not get to the bottom:*
> *death is a hole*
> *in which we are all buried*
> *Gentile and Jew*
>
> *The flower dies down*
> *and rots away*
> *But there is a hole*
> *in the bottom of the bag*
>
> *It is the imagination*
> *which cannot be fathomed*
> *It is through this hole*
> *we escape*

THE WOODS HOLE CANTATA

Through this hole
at the bottom of the cavern
of death, the imagination
escapes intact.

No believer in an afterlife, Williams had for over forty years put his bet on his power of invention. His goal was the forging of a new, distinctly American poetic "line." This line was to call up images of the everyday world in the patterns of real speech—what Williams called the "roar of the present"—which would separate our poetry from myth, metaphor, and allusion, from "the past above and the future below." He insisted that he was not only after images ". . . as some thought, but after line: the poetic line and our hopes for its recovery in the sense that one recovers a salt from solution by chemical action." In such an endeavor, words were to be used as pieces of type, elements of design, dispersed on the printed page like paint squirted on an unprimed canvas. To realize these complementary goals, Williams was able to draw on two cultural strains which constituted his personal history, and which framed a sort of dialectic.

In his objectivist, realistic mode, Williams paid tribute to his major predecessor at the task of uncluttering the language: Walt Whitman of Camden. But there were no lilacs in the dooryards of twentieth-century Passaic. Williams took his cues from the calls of his immigrant neighbors (*Come on! Wassa ma'? You got broken leg?*), from the cadences of the police docket (*I think he means to kill me, I don't know what to do. He comes in after midnight, I pretend to be asleep*) and from the rhetoric of American progressivism

(. . . *I refuse to get excited over the cry, Communist! they use to blind us*). He was closely allied in this effort to the social realists, and especially the precisionist painters whom he knew so well: Charles Sheeler, Ben Shahn, Louis Lozowick. The work of these artists paralleled the literary efforts of the objectivists, whose press first published Williams' poems in 1934. This objectivist, descriptive strain of poetry looks as clean—as black and white—on the page as a Lozowick lithograph.

From *Paterson 1:*

> *Things, things unmentionable*
> *the sink with the waste farina in it and*
> *lumps of rancid meat, milk-bottle tops: have*
> *here a tranquility and loveliness*
> *Have here (in his thoughts)*
> *a complement tranquil and chaste.*

But Williams was not only a precise American realist, who turned things unmentionable into the stuff of tranquility. His new poetic line carried the marks of a more complex—and a more cosmopolitan—strain than any to which Williams readily confessed. He may have waged a constant war against the professional Anglophiles of English departments, he may have called T. S. Eliot's *The Waste Land* "the great catastrophe to our letters," but he was by no means an autodidact or small-town primitive. "Pop never in his life made more than the barest possible income . . . Yet we did have an occasional case of Château Lafite in the cellar," he announced in his autobiography.

Pop was English, never acquired American citizenship, and traveled a great deal of the time in South America as a salesman of pharmaceuticals. Williams' mother was born in Mayaguez, Puerto Rico, to a French mother and a Sephardic Jewish merchant. His maternal uncle, Carlos—whose name was to give Bill Williams the touch of the poetic—was a Paris-trained physician who practiced in Port-au-Prince. William Carlos Williams was no product of the local schools of New Jersey. He had spent a childhood year at an exclusive Swiss school near Geneva, commuted across the Hudson to prep at Manhattan's experimental Horace Mann, and made the varsity fencing team at the University of Pennsylvania. It was at Penn that he fell in with the poets Ezra Pound and Hilda Doolittle (H.D.).

After house staff training at French and Babies' hospitals in New York, he did postgraduate work in pediatrics at Leipzig and traveled extensively in Europe. In England, Williams met Yeats in the company of Ezra Pound, who by that time had already established himself as a poetic prodigy. In Italy, he stayed with his brother Ed at the Villa Mirafiori of the American Academy, where Ed, a graduate of MIT, had won a Prix de Rome. Then back to Rutherford, where he not only plunged into practice, but into the life of the avant-garde of Greenwich Village as it caught European fire from the Armory show of 1913. There Williams was pursued through the streets by a mad German baroness, and visited salons which featured Nancy Cunard (of the ocean liners) and Vladimir Mayakovsky (of the Russian revolution). Williams spoke French and Spanish, translated from both languages, and could pass in German. All in all, this was hardly the curriculum vitae of your average suburban practitioner.

When his poetic line drew on this cultural strain of cos-

mopolitan energy, it carried with it echoes of "impression-
ism, dadaism, [and] surrealism applied to both painting
and poem." The line was fractured on the page in the
manner of collages, there were wild, sometimes incantory
rhythms that might have derived from the lifelong spiri-
tualism of his mother and maternal grandmother (seances
in the living room!). In this continental mode, his verse
ran a course parallel to the work of yet another group of
painters Williams admired greatly: Francis Picabia, Juan
Gris, and Pawel Tchelitchew. This piece, from *Paterson 5,*
displays the two strains of Williams' muse:

> *Satyrs dance!*
>   *all the deformities take wing*
>   *Centaurs*
> *leading to the rout of the vocables*
>   *in the writings*
> *of Gertrude*
>   *Stein—but*
>   *you cannot be*
> *an artist*
>   *by mere ineptitude*
> *The dream*
>   *is in the pursuit!*
> *The neat figures of*
>   *Paul Klee*
>   *fill the canvas*
> *but that*
>   *is not the work*
>   *of a child . . .*

I saw love
> *mounted naked on a horse*
>> *on a swan*
the tail of a fish
> *the bloodthirsty conger eel.*

Here is Williams doing all the things he does so well, reminding us of the work involved in setting simple words, like tesserae, on the page: this is not the work of a child! It is, in fact, the work of a mature artist who has both consciously and unconsciously grasped the two opposing strains of his nature and united them.

Williams had little use for popular psychoanalytic jargon, especially when applied to biographical or literary material. Nevertheless Robert Coles, introducing *The Doctor Stories,* recollects that Williams once compared the insights he obtained from his writings—the "descents into myself"—to the insights of the analytic experience. Riding that winter night in his car, chest aching with the fear of worse to come, was a doctor whose self-diagnoses were written in language that transcends the clinical. For Williams faced head on, in poetry and prose, the dynamics of a common enough family romance. In his autobiography he tells us of the last dream he had of his father. It is 1918, shortly after his father's death; the old man appears on the staircase of his New York office building, carrying business letters. Turning to his son, the father tells him, "You know all that poetry you're writing? Well, it's no good." Williams recalls that he awoke trembling, and in a phrase reminiscent of Dante, assures us that "I never dreamed of him since." Echoes of that bitter encounter can be heard in Williams' furious rejections of Eliot—that other Anglo-

American—and eventually of Pound (who, in the wordplay of their irascible correspondence, finally became devalued).

His mother, on the other hand, played both muse and audience to his craft. In "The Artist" he describes the narrator, the poet-dancer, performing a marvelous ballet leap in the house: the wife looks in from the kitchen to ask "What's going on here?", but the mother cries "Bravo!" in surprise, and claps her hands. In life and art Williams seemed to have resolved the dialectic between the factual, English, prosaic aspects of his father's persona and the less rational, French, and artistic aspects of his mother's. He became both doctor and poet, suburban *pater familias* and Lothario of the studios. His poetry, too, resolved the dialectic between objectivism and surrealism, between American fact and French fancy, between patriotism and social protest. These resolutions came by way of his unique capacity for invention. And, in the manner of all great art, the poet's inventive effort leaves sweat on the brow of his reader. From *Paterson 2:*

> *Without invention nothing is well spaced*
> *unless the mind change, unless*
> *the stars are new measured, according*
> *to their relative positions, the*
> *line will not change, the necessity*
> *will not matriculate: unless there is*
> *a new mind there cannot be a new*
> *line, the old will go on*
> *repeating itself with recurring*
> *deadliness: . . .*

The invention of this sort of line—which was to be his personal achievement in verse—seems to me to have a unique, if perhaps trivial, base in his daily life. His son, Dr. William Eric Williams, made available some of his father's clinical records and notes; these have been reproduced in the William Carlos Williams Commemorative Issue of the *Journal of the Medical Society of New Jersey*. One can find there a copy of one of the typical medical charts of Passaic General Hospital, in which the case is detailed of a five-week-old baby seen by Williams in December of 1946. These sheets, on which he must have written daily, are divided by a line ruled down the middle of the chart. This compositional element, which was by no means a common feature of the medical records of the time, forces the physician/narrator to break his notes into short lines of two to three words. Thus, in the record illustrated, Williams' notes seem broken into the cadence of poetry.

> *The child has not done*
> *well with its feedings*
> *since birth*
>
> *Breast feeding was*
> *not successful*
>
> *On a reasonable formula*
> *of milk, water and D.M. #1*
> *a diarrhea developed*
> *1 week ago*
>
> *Ears examined*
> *the right found*
> *to be red and bulging*
>
> *Sent to Dr. Wm Schwartz*
> *to be opened.*

The middle ear infection
was thought to be
the root of the problem.

Compare the final diagnosis on this chart with the final diagnosis in the poem called "Proletarian Portrait"; both follow a chain of clinical reasoning jotted down in chart-like fashion:

A big young bareheaded woman
in an apron

Her hair slicked back standing
on the street

One stockinged foot toeing
the sidewalk

Her shoe in her hand. Looking
intently into it

She pulls out the paper insole
to find the nail

That has been hurting her

In his most eloquent testament to his medical work, a chapter of the autobiography entitled "The Practice," Williams confesses what he had learned from the years of treating sick babies, their mothers, the poor, from the thousands of house calls. He discovered that the voices of these patients, these people, had given him access to the sounds of his personal poetics. They were offering the doctor, in their own voices and with their own bodies, a profound story of the self, a poem:

The girl who comes to me breathless, staggering into my office, in her underwear a still breathing infant, asking to lock her mother out of the room, the man whose mind is gone—all of them finally say the same thing . . . For under that language, under that language to which we have been listening all our lives a new, a more profound language, underlying *all the dialectics* [my italics] offers itself. It is what they call poetry.

Williams made it back safely to Rutherford that night, which he spent in intermittent pain. (Did he wake Floss?) Only after he rose and breakfasted, however, did his pain become unbearable. He called his cardiologist, Dr. Gold, who diagnosed a small anterior myocardial infarction. In those days, before the days of coronary care units, he was appropriately treated by six weeks of bed rest.

Relieved of medical work, Williams went on a remarkable spree of writing and publishing. He started on his autobiography, plunged deeply into Books Three and Four of *Paterson,* and, indeed, in the space of the next five years published many of the books by which we now remember him. These include the *Collected Later Poems* (1940–1950), and *Make Light Of It: The Collected Short Stories* (1950). Honors and recognition finally came his way—as if to compensate him for those long years in his thirties, forties, and fifties when Ezra Pound accused him of "pissing his life away." While still convalescing from his infarction, he was informed that he was to share the Russel Loines Award of the American Academy of Arts and Letters with Allen Tate. In 1950, he received the first National Book Award for poetry, for *Paterson 3.* And in 1953, he shared the Bollingen Prize—that prize which was first won by Pound in a sea of controversy—with Archibald MacLeish.

But the poet's life seems to have been played out to the moral tune of the old Spanish proverb "You get what you want from this world, but you pay for it!" At the peak of the recognition for which he struggled—the lectureships, the critical praise, the ardent disciples—his strength and energies were sapped by a series of strokes which affected not only body but mind. With pluck and tenacity, he learned to peck out verse with one hand, and wrote so well in the last decade allotted to him (1953–1963) that he was posthumously awarded both the Pulitzer Prize and the gold medal for poetry of the American Academy of Arts and Letters.

Williams lives. The bookshops have his works in print, the museums are full of his friends Demuth, Sheeler, Lozowick. There is a WCW newsletter. The courses of medical humanities are filled with students who read him for his subject matter alone. And in these times when androgyny is recapitulating phylogeny, many of his readers are delighted by his celebrations of heterosexual love. But the Williams who will outlast this flurry of attention is the poet of urban invention, who described the American landscape in a series of keen images which he set to a music through which we can hear the roar of the present. Unlike Pound and Eliot, he had no hankering for the past of empire and the church. Williams knew that

> *—the times are not heroic*
> *since then*
> *but they are cleaner*
> *and freer of disease.*

By paying attention to *things,* those bytes of clinical observation which make up the life of a people, rather than to abstract *ideas,* which in his lifetime had caused so much mischief, Williams came to terms not only with his own poetry, but with ours.

> —*Say it, no ideas but in things—*
> *nothing but the blank faces of the houses*
> *and cylindrical trees. . . .*
> *Say it! no ideas but in things. Mr.*
> *Paterson has gone away*
> *to rest and write. . . .*

Couching the Question:
Freud Revisited

SINCE my student days at Bellevue, I have never enjoyed rounds on the prison ward. Over the decades, buildings have come and gone, corridors have vanished and elevators become automated, but the prison facility is still inconveniently separated from the general medical service by a maze of hallways and an elevator ride or two. The ritual of access has remained a nuisance. Our troop in white—students, house officers, attending physicians—waits outside the ceiling-high gate for a correction officer to drop his *Daily News* to let us in. We have ample time to consider the implications of the notice stenciled over the lock:

> ALL WEAPON'S {sic} MUST BE SURRENDERED
> PRIOR TO ENTERING PRISON WARDS.
> USE DESIGNATED AREAS.

We next enter an anteroom, where we sign our names and destinations in a ledger monitored by a second officer

in blue. Through a bulletproof window we see a third officer locked in a kind of nursing station; we can obtain information from him by means of a two-way microphone
which perforates the glass. A second set of gates is unlocked for us as we enter the male ward, in which a dozen
or so prisoners are dispersed among four-bed sickrooms.
Reflecting the population of the Riker's Island penitentiary
from where they have been transferred, the men are young
and mainly black or Hispanic. Many have intravenous tubes
hooked into remnants of veins as they lounge before television sets tuned to soaps. Sick or convalescent, they look
remarkably muscular as the result of hours devoted to body-
building behind bars. Their arms bear tattoos of crimson
and blue: graffiti of the skin. The rooms send signals of
acne and testosterone, they remind me of squad rooms in
basic training at Fort Dix.

At the foot of Mr. Brown's bed, I am handed his chart
by the resident; the problem seems to be an almost routine
case of gonococcal arthritis. The patient's knee has been hot,
swollen, and painful for three days. Fluid taken from the
joint was filled with the white cells of bacterial inflammation, his fever had been moderate, and a history has been
obtained of frequent sex with many women in the months
before his confinement to Riker's Island a fortnight ago.
Awaiting trial, he had complained of urethral "strain."

History and physical examination completed, our cohort
moves to the nursing station where we look at the X-rays
which show only soft-tissue swelling of the knee. I casually
asked the resident what had caused Mr. Brown's arraign-
ment.

"I guess they got him on drugs. He has needle tracks
and what sounds like a vague history of hepatitis last year.

He's been here before, but I couldn't find the old chart" is
the reply.

"Did you ask him?"

"Not directly."

We glance over the chart. It contains the usual details
obtained by medical student, intern and resident: age, previous illness, review of systems, family history, schooling.
Nowhere is recorded the alleged offense for which he was
imprisoned. Curious as to whether Mr. Brown's joint complaints might relate to possible liver disease, we revert to
the bedside to ask the patient why he was arraigned. We
reason that it would be difficult for a patient with chronic,
active hepatitis to hustle for drug money on the streets. He
is not very forthcoming:

"A family thing. They're trying to pin it on me. Drugs
is just an excuse."

We ask a few more questions about his health—if he has
become tired easily, his urine has turned dark, he has lost
weight or had fever, whether other joints have hurt him,
if he has lost his taste for cigarettes. He is clear and precise
in his answers to these questions: there is nothing in the
recent past to suggest liver disease. The chemistries on the
chart support this.

Agreeing on the probable diagnosis of gonococcal arthritis, we file out to the anteroom. We stop to ask the
officer behind the glass what the charge is against Mr.
Brown.

"Oh, that one," comes the voice over the microphone.
"They say he raped his daughter by his common-law wife
and nearly beat the girl to death. She was nine. I know all
about him from the guys what brought him. High bail.
Case won't come up for a while."

We traipse back to a conference room in the main hospital, where, over coffee, we discuss the management of Mr. Brown's knee. But I'm bothered about Brown, the alleged rapist. I wonder out loud why we have this selective reluctance to ask prisoners just what the offense is of which they are accused or convicted. For it has been my experience over three decades that while we do not hesitate to ask patients in the prison ward to describe the most intimate details of their personal lives, we seem to have declared the topic of criminal charges off limits. We obtain answers to other questions: whether their houses are infested with rats, if they pass urine in their sleep, have pain on intercourse, practice anal sex with man or beast; we determine their choice of drug and how many dollars are required daily to support their habit. We avoid only that question which would first engage the curious layman.

"I guess it must be unconscious," says a bright resident. "I think we're afraid. I mean after you've gone and examined this sick guy, and been up with him half the night pulling him around, you sort of get to know him. He turns out to be not too different from the other patients: he just got caught."

"I don't understand," I reply. "If that's true, then why are you afraid of him?"

"I think I'm afraid for myself. There's some criminal element in all of us, in our subconscious, and I suppose I sort of identify with him because of that. You know: there but for the grace of God and my parents' money . . ."

A medical student chimes in:

"But I sure don't identify with all of them: some of them really scare the shit out of me. When the cops come into the emergency room with some huge hulk with razor cuts

on him, I'm not going to ask him how he spends his evenings and what the cops got him for. I'm *afraid* to know,
and, besides, I don't want to get him angry."

"There's another view," says one of the fellows. "I think
of what life is like for anybody in jail. I identify with these
guys also, but when I put myself into their place I just
worry about being raped. There's all that homosexual attack going on all the time . . . I think that deep down I
don't want to hear about any of that."

One of the women says that can't apply to her:

"Oh, I think I don't ask them because these people make
me feel guilty. They're poor and victimized. They may
commit crimes, and I'm glad they're punished, but they
all come from bad homes without real fathers, and they are
mainly from minorities and can't get good jobs. I feel guilty
and that makes me afraid, too. That poor drug-crazed fellow couldn't help what he did. He overdosed on the family
romance, if he actually did it."

"Well, what about that?" I ask. "Would it make a difference to you whether he had or had not raped his daughter? I mean, in the way you talk to him or treat him?"

"I know it shouldn't, but it would," says the first resident. "But of course we'll never know; he'll be out of
Bellevue before the trial, *if* he comes to trial. It doesn't
matter. He looks as if he *could* have done it: any of these
guys could have done it, they're so used to violence. They
act out."

"He probably did it, and if he gets convicted it's because the D.A. was good and if he gets off it's because his
lawyer was better," says the woman. "I'm sorry we found
out what he did."

"But can we really ever be sure whether he did it or not?"

asks the student. "We would have to sit through the trial, and even then, how can you trust a nine-year-old? You sure can't ask the prisoner; these guys are habitual liars. I say it doesn't matter whether he is legally guilty or not."

"That's right," rejoins the woman resident. "We always assume that if he is a prisoner then he is probably guilty but that should be none of our business, we ought to treat him for his knee, and send him back to Riker's."

"But what about our obligation to treat the whole patient? If his addiction and his venereal disease are really social problems that made him act out, shouldn't we really know if . . ."

The discussion continues a few minutes more as a mixture of anxious ventilation, condensed psychiatry, and honest self-examination. We adjourn to the daily routine of the hospital and I forget the topic of child rape until I pick up *The New Yorker* that evening after dinner.

The magazine contains the first of a two-part series by Janet Malcolm on the problematical career of Jeffrey Moussaieff Masson, a series that is to become a salon scandal in Manhattan analytic circles. And from among the sleek pages of the journal I hear faint echoes of the morning's discussion couched in terms that might make little sense to Mr. Brown.

The problem, in brief, centers on Freud's clinical decision as to whether fathers actually seduce their daughters or whether the daughters fantasize the act. The Malcolm articles describe the rise and fall of Masson's career in the high circles of orthodox Freudian analysis. A Harvard-trained professor of Sanskrit, Masson had turned to psychoanalysis. His verve, scholarship, and devotion to Freud so impressed Kurt Eissler, the most influential of the trustees and Secretary of the Freud Archives, that Masson was hired as Projects Director of the Archives, and slated to succeed Eissler himself as high priest of the sect. The Freud Archives include hundreds of letters deposited partly with the Library of Congress, some of which will not be available for general inspection until the twenty-first century. But the scandal arose when the heir apparent was set loose in the archives and given access to Anna Freud's repository in London. It seems that Masson's study of the material resulted in what amounted to a major discovery—or so it appeared to Masson—which he outlined at a now-notorious lecture to a meeting of analysts at Yale in June of 1981.

Masson produced quotations from letters of Freud to his rhinological confidant, Fliess, which suggested to Masson that Freud had trimmed the sails of psychoanalytic theory to the current winds of sexual prudery. For in his earliest psychoanalytical inquiries, Freud had become convinced that several of his women patients had suffered sexual trauma in childhood consequent to seduction by their fathers: rape, more or less. To these traumas he traced the roots of adult neuroses. In later papers, however, Freud retracted his earlier views and denied the basis in fact of incest recalled on the analytic couch, suggesting that the women had fantasized scenes of seduction. The genesis of these fantasies was first described by Freud in his 1915–17 lectures at the University of Vienna. It may be worth noting that. he does not, indeed, raise the Oedipus complex in front of this audience but relates the fantasy to guilt over masturbation:

> . . . when girls who bring forward this event in the story of their childhood fairly regularly introduce the father as the seducer, neither the phantastic character of this accusation nor the motive actuating it can be doubted. When

no seduction has occurred, the phantasy is usually employed to cover the childhood period of auto-erotic sexual activity; the child evades feelings of shame about onanism by retrospectively attributing in phantasy a desired object to the earliest period.

In his Yale lecture, and his subsequent book *The Assault on Truth,* Masson attributes Freud's shift of view to a conscious bending of the truth. Masson claims that Freud transposed the charged subject of actual child seduction to the realm of fantasy in order to lessen the resistance to psychoanalysis (and its practitioners) of Viennese physicians in whose power it was to grant academic respect to the field. Masson adduces evidence that, contrary to Freud's earlier diagnosis of child abuse as the causative trauma in his patients, Freud's later revision of theory was a willfully dishonest act perpetrated in the service of the general acceptance of psychoanalytic technique. Masson's accusations suggest that Freudian theory is invalidated by this act of obsequy: if inner fantasy can replace external reality so easily, where is the science? Masson's grandiose comment on his new insights into the seduction theory were reported in *The New York Times:* he was certain that this evidence of Freud's duplicity shattered the foundation of analytic therapy. "They would have to recall every patient since 1901. It would be like the Pinto."

In response to this apostatic view, Masson was removed from his position as Project Director of the archives and shunned by keepers of the Freudian flame. In the Malcolm account the venerable Kurt Eissler, most eminent and orthodox of living Freudian disciples, is cast in the role of a father betrayed by a spiteful son; her articles recount the family feud with insight and compassion.

It is not my purpose to interpret the motives of Freud, the documents presented by Masson, nor to speculate on the inner dynamics which motivated the central figures caught by this tempest in an analytical teapot; such matters are beyond my competence. What strikes me as remarkable is that *any* field of clinical inquiry should be perceived to be vulnerable to charges such as those which Masson leveled against Freud. Could the Freudian canon rest upon so fragile a base? Psychoanalysis has indeed had a rough time in medicine lately, battered as it is by advances in psychopharmacology, the powerful tools of neurobiology, and new appreciation of the unique social setting of the case histories described by Freud. Critics point out that turn-of-the-century Vienna may not be the best template for patterns of family behavior at other times and in other places: the ghettos of American cities, for example. Moreover, rigorous medical scientists, such as Peter Medawar, have argued cogently that the clinically based hypotheses of Freudian psychoanalysis—which cannot by definition be disproved—do not meet the first criterion of scientific statement. The result of this disenchantment with Freudian technique on the part of the medical profession, which had grudgingly and belatedly integrated Freudian insights into the schools only a short while ago, is that the best and the brightest of our graduates with a bent toward psychiatry are newly enraptured by the organic determinants of behavior. Their grant applications deal with neurons and receptors, peptides and nucleic acids, *Aplysia* and fruit flies.

Freud would not really have objected. His career began in experimental neurophysiology, and in his *Interpretation of Dreams* (1899) he hoped that deeper research will one day trace the path further and find an organic basis for the mental event . . . [and] find a means of picturing the movements that accompany excitation of neurons.

Shades of the PET scan! Indeed Freud was sufficiently aware of the genetics of his time, and influenced by August Weismann's discovery of the continuity of the germ-plasm to anticipate the modern hypothesis of the selfish gene. In *On Narcissism: An Introduction* (1914), Freud suggested that the individual

> . . . is only an appendage to his germ-plasm, to which he lends his energies, taking in return his toll of pleasure—the mortal vehicle of a (possible) immortal substance . . .

Nevertheless, to most of our students, Freudian theory and practice seem fussy, old-fashioned, and untrustworthy. When critics like Masson put the veracity of the founder himself to question, it may well appear to students that the whole psychoanalytic enterprise is at risk for intellectual bankruptcy. Freud has clearly described these high stakes in his first lectures to a general medical audience:

> It is impossible, therefore, for you to be actually present during a psycho-analytic treatment; you can only be told about it, and can learn psycho-analysis, in the strictest sense of the word, only by hear-say. This tuition at second hand, so to say, puts you in a very unusual and difficult position as regards forming your own judgment on the subject, which will therefore largely depend on the reliance you can place on the informant.

If, as Masson has endeavored to prove, the informant is unreliable, then Freudian theory and practice cease to matter. Indeed, Masson provides some evidence that Freud suppressed his own knowledge of actual cases of child rape or assault. The argument then, would appear to hinge upon the kind of questions that arose at the bedside of Mr. Brown: Did he or did he not actually assault his nine-year-old daughter? Ten years from now, what sort of story would be elicited from her by an analyst?

I would submit that the entire line of argument is irrelevant to the larger question of the validity of Freud's work. Clinical observations made on a limited, selected group of patients—and psychoanalytic practice dictates that the number of patients studied in a lifetime of practice is necessarily limited and selected—cannot yield statistically useful data as to prevalence of any condition or state. Indeed, when we look at the qualitative terms used by Freud to describe his own experience, we sense his honest effort to make sense of such material:

> The phantasy of seduction has special interest, because *only too often,* it is no phantasy but a real remembrance; fortunately, however, it is still *not as often* real as it seemed at first from the results of analysis [my italics].

How can one argue the distinction between "only too often" and "not as often"? Of what importance to the survival of psychoanalysis is it whether 25 or 35 percent of recollected seduction—or rape—is real or the product of fantasy? Child abuse and incest probably vary in incidence from culture to culture but no analyst can report general epidemiology from his office. It is to Freud's credit, I should have thought, to have pointed out both the reality and the fantasy. It may perhaps have been a form of hubris for Freud to have been certain that he could distinguish one from the other on the basis of analytically induced recollection, but even that seems to miss the point of his contributions. For surely what Freud discovered was the process whereby a language could be learned with which he could describe the deepest sources of behavior. Before Freud, psychiatric discourse contains no reference to the world of the *dynamic* unconscious: that vast, archaeologic site of personal history in which the shards of fantasy and reality lie buried together.

Indeed, internal evidence suggests that over the course of Freud's career, while the analytic process remained more or less constant, his interpretation of unconscious mechanisms underwent major revisions. It is difficult to imagine that all of these revisions were due to willful deceit. Three examples should suffice.

First, in his account of the "Wolf-Man" (1918), Freud describes his patient's childhood recollection of the "primal scene" when on a hot summer afternoon he watched his parents copulate—not once, but thrice, and in the manner Freud delicately describes as *more ferarum*—so that the child could see the genitals of his parents. Later in the analysis, Freud was led to the tentative hypothesis that the "Wolf-Man" had, in fantasy, displaced his childhood observation of the manner in which dogs copulate to a primal scene enacted by his parents. But at the end of the monograph, reconsidering the interim interpretation, Freud concluded that "the effects that followed from this in the patient's life can most naturally and completely be explained if we consider that the primal scene, *which may in other cases be a fantasy* [my italics], was a reality in the present one."

Second, Freud, having discovered the phenomenon of repression, was for a long time of the opinion that this psychic mechanism for handling charged, conflicted material provokes anxiety. But by 1927, in *Inhibition, Symptoms and Anxiety* he was forced by clinical experience to confess, "It is not pleasant to think of it, but there is no use in denying that although I have repeatedly put forward the thesis [that repression provokes anxiety], this statement of mine is not incorrect, but superficial." Indeed Freud reverses his field to conclude that, in the main, anxiety induces repression.

Third, in *Totem and Taboo* (1913) Freud attributed the origins of religion to the universal, subconscious machinery of Oedipal dynamics. But by 1927, in *The Future of an Illusion,* he discarded this early view as too simplified, and replaced it with the more conventional notion that religions arose to meet general human feelings of weakness and helplessness.

More of these examples exist. They give evidence that Freud's method was honestly based on the principles of clinical investigation: he gathered data from patients by analytical methods; formulated hypotheses to account for the material revealed in his practice; and modified earlier interpretations as new "evidence" was produced from the laboratory of the couch. Is that an assault on truth? (The story of this evolution in Freud's thought as revealed by shifts in language, metaphor and other literary devices is well described by S. E. Hyman in *The Tangled Bank.*)

It is the language of Freud, the process of his discourse with the unconscious, that remains with us today. Like it or not, that language shapes the way we interpret our behavior to ourselves and to others. We have no alternative. Remember that discussion at Bellevue about our patients on the prison ward? What were the words we used and why did we use them? One can dismiss as learned responses, as pseudoanalytic jargon, our references to the "unconscious," to "identification," to "guilt," and to "acting out." But the fact remains that this generation of young doctors is more in touch with their own feelings, their own psychic processes and those of their patients, than were the Viennese physicians whom Freud addressed in 1915. And that difference, I daresay, is in no small measure due to the impact of Freud and the psychoanalytic movement.

The heroic generation of European analysts (Freud, Breuer, Ferenczi, Hartmann, Kris, Rado, Reich, Reik, *et al.*) has dictated the very questions we ask about behavior. It is useful to remember that there have been fewer psychoanalytically trained psychiatrists than evolutionary or molecular biologists, atomic physicists, or Marxists: each of these groups has had a comparable impact on the modern world. The discourse of the Freudians has not only changed the purview of medical psychiatry, but enlarged the scope of the social sciences and the arts. Freudian language and images bestride the humanities like a colossus: from Dali to Balthus, Cocteau to O'Neill, Kafka to Borges, Lang to Hitchcock, Trilling to Barthes. It may also be useful to remember that the Freudian revolution is based neither on bare rhetoric nor on *a priori* reasoning but on painfully acquired insights into clinical material.

To quote Malcolm: "When Freud dropped the seduction theory and introduced the theories of infantile sexuality and the Oedipus complex, he transformed psychoanalysis from a form of social psychiatry into a depth psychology."

A challenge remains for Freud's successors. Their task will be again to frame a language of social or biological psychiatry that permits us to understand not only personal neuroses, but also the social offenses of Bellevue prisoners. That language will have to be as convincing as that of the depth psychology, which Freudians would claim is common to convict and house officer. I am persuaded that the questions these successors will ask in the course of that endeavor will, perforce, be influenced by the Freudian description of our unconscious lives. Is this description based on willful deceit, as Masson suggests? Or did Freud honestly modify his seduction theory on his way to another discovery? We have firm evidence for the second alternative. Indeed Freud suggested (1925) that, although not in every patient, surely in some, many neurotic symptoms

> were not related directly to actual events but to wishful fantasies; as far as the neurosis was concerned psychical reality was of more importance than material reality. I do not believe even now that I forced the seduction fantasies on my patients, that I had "suggested" them. I had in fact stumbled for the first time upon the Oedipus complex, which was later to assume such an overwhelming importance, but which I did not recognize as yet in its disguise of fantasy.

It was his modification of the seduction theory—from its basis in reality to its origin in fantasy—that made possible his discovery of the Oedipus complex. The process forged a new language.

Sixty years later, Freudian discourse has not been replaced by any other coherent language with which to describe our inner life. We are incapable of discussing our deepest feelings about each other and ourselves without reversion to modes of description first employed by the psychoanalysts. Contemporary psychiatry has certainly progressed far from the relatively rigid perimeters of the

Freudians, but it is undeniable that it was they who launched the assault of reason on the barriers of personal angst. There may be no such *things* as the "ego," "superego," or "id"; the Oedipus complex may not be the universal explanation for disorders of the family constellation; and analytic theories of the etiology of schizophrenia are almost meaningless these days. No matter. Masson may wax apoplectic over Freud's "dishonesty," other moles in the earthworks of the Archives may discover that Freud was having it off with Minna Bernays. No matter. These noises are eclipsed by the clear music of Freudian canon. The message is the *process:* the rational interpretation of what we choose to say or not to say, to recall or not to recall, to dream or not to dream. I do not know what awful secrets will be unlocked by the Library of Congress in 2102 when the last letters of the Freud Archives are made available to scholars. If the Library of Congress, and scholarship itself, survive to 2102, their persistence may well be due to answered questions. These will be questions we have asked of our most secret and violent selves, and in all likelihood will be couched in terms of Freudian discourse.

No Frogs in Berlin

THE lights of Berlin crept under the wing as our British Airways 727 circled toward a landing on an autumn night in 1983. In the cabin, seat trays were stowed and safety belts fastened. As the no-smoking lights flashed, I tucked notes and reprints into my carry-on, and mentally rehearsed an outline of testimony I was scheduled to give the next morning. This was my first visit to Berlin, where I was about to serve as an expert witness on the safety of nonsteroidal anti-inflammatory drugs, aspirinlike drugs, used chiefly in the treatment of arthritis. The hearing was convened by the *Bundesgesundheitsamt,* or BGA, roughly the German equivalent of our FDA, and I had spent a good part of the short flight from London deciding whether to tell the BGA commissioners a curious tale of Australian frogs.

On many flights, aided no doubt by hypoxia and alcohol, I fill the minutes between descent and landing with active fantasies. These tend to be based on the black-and-white movies of my childhood. Indeed, when I boarded the plane at Heathrow, I fully expected to land at Berlin's Tegel Airport after an imaginary voyage among the ghosts of Bomber Command. Forty years ago, along this flight path, young Yanks in Fortresses and Englishmen in Lancasters left fields in East Anglia to struggle with volleys of flak: good guys out to slay the dragons of the Third Reich. But on this flight, the passengers were a microcosm of the less troubled generation of the Common Market. Young German businessmen amiably passed along small change and miniature liquor bottles from the flight attendants to British tourists. In consequence of this reassuring *bonhomie,* I was free to ruminate on the gastric brooding frogs of Queensland.

I had first heard about the frogs at one of our weekly journal clubs at NYU, where a scholarly clinician, Larry Faltz, presented the work of Michael Tyler. The report was unusual, not only in scope, but in the precision of its prose. It began with a cadence that was pure Conan Doyle.

> In southeast Queensland, there is a rare aquatic frog *Rheobatrachus silus.* The female frog swallows fertilized eggs or early stage larvae and broods the young in her stomach. The young frogs eventually emerge by way of her mouth.

Since the business of the stomach is to break down ingested foodstuffs by means of acid and pepsin, Tyler and his associates reasoned that the developing young must elaborate a substance that inhibits gastric secretion. In a series of ingenious and intricate experiments, the Australian zoologists isolated such a substance, and identified it as fatty material called *prostaglandin* E_2. They presented firm data which supported their contention that *R. silus,* but not other species of frogs, had "developed a mechanism whereby prostaglandin secreted by the larvae inhibits acid secretion in the stomach of the female until the larvae have completed development and emerged as juvenile frogs by way of the female's mouth."

Now this curious bit of zoology had direct bearing on my trip to Berlin. For it has been amply documented, since the work of England's John Vane in 1971, that nonsteroidal anti-inflammatory drugs not only inhibit the synthesis of prostaglandins, but that they exert their most serious side effects on the gastrointestinal tract. Indeed, among the major concerns of the BGA with respect to nonsteroidal drugs was their capacity to induce peptic ulcers and gastric hemorrhage. Moreover, a decade of research had demonstrated that prostaglandins not only inhibit gastric acid secretion, but also stimulate the secretion of gastric mucin, the protective barrier. When nonsteroidal anti-inflammatory drugs are given, the normal protective function of prostaglandins is removed, and the stomach is free to churn itself up. I resolved to bring up the frog model as a memorable experiment of nature which would make it clear that any effective inhibitor of prostaglandin synthesis would, by definition, possess the potential for inducing gastrointestinal irritation: it would produce ulcers.

The plane landed. It was ten-thirty in the evening and the airport was deserted except for a dingy platoon of locals waiting for the arriving passengers. The various airline counters were already shut for the night, lights were dim, and a few Balkan *Gastarbeiter* were at work with mop and broom. The movie scenario switched in my mind to the world of John Le Carré as a trench-coated chauffeur disentangled himself from the group at the customs gate to tuck me into the back of a sinister black Mercedes. The atmosphere of paperback espionage was not helped when, without hesitation, the driver opened conversation in German: someone had done his homework. I was informed as to which hotel I was quartered at and given an agenda for the morning; landmarks of the city were pointed out in short bursts of Berliner staccato. We drove through empty, clean boulevards to the Kurfürstendam, skirting the gloomy tower of the war-damaged Memorial Church. The limo rumbled past modern shopping streets in which the usual gaggle of international chain stores (St. Laurent, Benetton, Lancel) looked at home among buildings that resembled the malls of surburban Atlanta. In response to the Berlin of Hitler, with assistance of the Red Army, the RAF and Eighth Air Force had effectively turned the city of Frederick and Bismarck into an duty-free airport shop with streets. In the distance, lit by floodlights which extended to the Wall, was the Brandenburg Gate, a triumphal arch which marked the boundary of East and West. It seemed possible that the mark of Cain was the just reward of unreason. Outside the hotel, several billboards were covered with graffiti which announced that the U.S. = S.S., apparently directed against the stationment of intermediate-range nuclear missiles in West Germany.

Inside the hotel, I found my room, arranged my slides and papers for the morning, twiddled the television for a few minutes, and finally prepared for the broken sleep of international travel. In the city where Stricker first reported that rheumatic fever could be successfully treated "*tuto, cito, et jucunde*" with the first aspirinlike drug, salicylic acid (1876), I had dreams of frogs and aspirin.

The hearing of the BGA was held in a vulgar version of

that generic Congress Hall which has recently sprouted like a wart in our major cities. A mixture of silver griddle cakes formed a massive facade; bursts of neon and black plastic tiles constituted the interior decor. This gargantuan edifice was fronted by a Prussian variation on the theme of heroic municipal sculpture; a thirty-foot-high *Niebelung* was clearly posing for the lead in *The Empire Strikes Back*. The hearing chamber was set up in the fashion of a cozy surgical amphitheater: the presiding commissioner and his aides were in the oval pit. In the audience of 250 or so were members of the press, witnesses from several German medical associations, assorted pharmacologists, representatives of consumer groups, batteries of scientists from industry supported by their administrative corps, expert German rheumatologists and clinicians, and a few foreign experts from Sweden, England, and the United States.

Leading the inquiry was Professor Dr. K. K. Überla, head of the BGA. Dr. Überla, a personable, crisp German of the technocratic school, looked as telegenic as any local anchorman. Nevertheless, he proved to be a cultivated, theoretical physician who had, indeed, written eloquently on science and discovery. Politically somewhat left of center, he has in his writings defended the value of untrammeled scientific research, paying appropriate tributes to the contribution of "breakthrough technology" to patient care. Indeed, in a recent symposium, he laid out his ground rules for the social assimilation of new technology.

> Better patient care clearly follows from progressive scientific breakthroughs: these cannot be exactly planned. The unexpected part of such progress is part of the scientific method itself and serves to optimize it. Indeed this progress is the natural outgrowth of the human capacity for ad-

aptation. But the growing scientific/technical culture has given rise to human problems. These problems, however, are not inherent in the nature of scientific/technical progress, but in the attitudes it engenders and in the social development of its discoveries. For the development of a technical, medical culture we can rely neither on scientific knowledge alone nor on philosophy alone; we need both to forge a culture out of the process of discovery.

This enlightened passage became particularly appropriate as the hearing over which Überla presided began to unfold. For the breakthrough discovery in this case was John Vane's finding that aspirinlike drugs inhibited prostaglandin synthesis. There seemed to be a very good correlation between the anti-inflammatory actions of known nonsteroidal agents (e.g., aspirin, indomethacin) and their capacity to inhibit the enzyme (cyclooxygenase) which formed prostaglandins. For over a decade now, drug companies have engaged in fierce competition to find other cyclooxygenase inhibitors which would be useful in the clinic. The results are staggering: threescore such drugs are now at various stages of clinical use somewhere in the world. In the United States, a good number have passed the standards of the FDA; among them are such drugs as ibuprofen, naproxen, tolectin, sulindac, diflunisal, fenoprofen, meclofenamate, and piroxicam.

Germany has several others, and in many formulations these drugs are given in combination with other ingredients. The development of this pharmacopoeia, and the intensive research this entailed, gave rise to new, unexpected discoveries, in a fashion described by Dr. Überla. It turned out that the "prostaglandins" known to Vane in 1971 did not necessarily mediate all aspects of inflammation, and by

1984, it is no longer entirely clear that nonsteroidals diminish pain and inflammation in arthritis exclusively by inhibiting prostaglandin biosynthesis. What has become evident, however, is that these drugs cause significant gastrointestinal side effects. Indeed, the capacity of nonsteroidals to irritate the gut, leading to ulcers, bleeding, and occasionally fatal bleeding, constitutes the basis of the "human problem" addressed by Überla in his essay and which prompted this hearing.

The BGA experts and representatives of German medical societies presented evidence that there had been a significant increase in serious gastrointestinal side effects as nonsteroidals became widely used, and argued that the availability of newer, long-acting drugs (such as the oxicams) increased the risk to which elderly patients were exposed. On their side, the pharmaceutical companies presented solid data which indicated that the newer drugs were all considerably safer to the stomach than aspirin, that proper usage and good medical practices could well have prevented those serious side effects which troubled clinicians and patients alike, and that the overall risk/benefit ratio was in their favor. The presentations dragged on *in extenso* on both sides. On the whole, the drug companies (chiefly American-based) presented their material in a statistical mode appropriate to a brief before our FDA. Their painstaking methodology, elaborated on multicolored slides, irritated Überla who, one eye on the journalists, complained in testy fashion: "After all, gentlemen, we in The Federal Republic are not exactly a scientifically underdeveloped country!"

An unbiased observer might have judged otherwise as a series of German experts in rheumatology, drawn in part from an older generation of spa doctors, recounted their

anecdotal experiences with the drugs in question. A refreshing exception to this local mode was a statement by
Kay Brune, a young professor of pharmacology from Erlan-
gen. Apologizing that his view of these drugs was colored
by his own laboratory experiments, he stoutly maintained
their general safety and utility. Experts from Germany and
abroad gave good reasons as to why there should be so many
drugs of this type: patients differ in their response, the
pharmacokinetics vary, etc. etc. The recital of data went
on: loquacity and pedantry so crowded the agenda that the
lunch break was canceled. Clusters of auditors filed in and
out at random, returning with half-eaten sausages from the
cafeteria.

As on airplanes, so in the course of slide talks, as the
lights dim so does my attention. Unconnected images came
to mind. I looked about the amphitheater; indirect lights,
deep shadows, microphone wires and earphones for simul-
taneous translation turned the auditors into Martians. Almost all of the scientists who represented drug companies,
and the outside experts, spoke in English. Our language,
if not our automobile industry, seemed to have emerged as
the residual victor of the European wars. The stocky, gray-
haired group of drowsy Germans in the front row reminded me of Stanley Kramer's *Judgment at Nuremberg*. Forty
years had made a gratifying difference, however. There too,
the proceedings had featured a discussion of drugs, but
happily the purpose of today's assembly was their safety and
not their misuse. On television the evening before, I had
caught fragments of film which showed young Berliners
crowding a demonstration against Yankee missiles. Pro-
testing at the same time against our interference with the
people of El Salvador and Lebanon, the demonstrators wore

old American combat jackets and jeans. I recalled films of other mass rallies in this city, and the image of John F. Kennedy standing at the Wall, proclaiming: *"Ich bin ein Berliner!"* (I am a Berliner). Irreverently, at the time, I was pleased that he had made no similar claim in Frankfurt.

The proceedings droned on. Agreement was evident on several points. Nonsteroidal anti-inflammatory agents of all makes seemed to be equally capable of provoking gastric irritation: stomach discomfort in 15 to 20 percent; discontinuance of drug because of these in 3 to 5 percent; serious adverse reactions in less than 0.5 percent. The drugs were all less toxic than aspirin, and just as effective. Many of the fatalities—and these were rare indeed—could have been avoided if doctors had prescribed properly. Perhaps the dosages should be modified in the aged, who might have trouble in excreting the drugs. The drugs were no cure for any arthritic disease, but did reduce inflammation; it was unwise to use them as part of the polypharmacy that is a therapeutic convention in Central Europe.

There was no significant discussion of uncertainties with respect to the mode of action of nonsteroidals; everyone agreed that they inhibited prostaglandin synthesis and went on to discuss their half-life in the body. Real questions were ignored, or raised only in passing. Is it worthwhile for doctor and patient to have available threescore drugs of the "me too" variety? Have the companies neglected the more difficult task of finding other kinds of antiarthritis drugs because of the fast dollars which can be made by mining the antiprostaglandin lode? Are the nonsteroidals, as a class, as safe as other widely prescribed drugs such as tranquilizers or antiulcer medications? Do prostaglandins cause inflammation? Does inhibition of prostaglandin synthesis explain the side effects but not the anti-inflammatory action of nonsteroidals? My own answers would probably have been *yes, no, yes, no* and *yes,* in that order. The Australian frogs would have been Exhibit One in response to the last question.

In the event, by the time direct comment was requested from me, I responded with a description of the usage of nonsteroidal agents by American rheumatologists, based on published experience. The data suggested that all drugs carry risks, and that this group of agents, in the hands of well-trained doctors, is probably among the least risky. Now as to their mode of action . . . I looked around to a sea of uninterested faces and slipped the frog slide back in my pocket. I was asked further questions as to the effects of nonsteroidals on liver, kidney, and skin, and fell into the general tone of long-winded explanation.

The discussion moved on. Despite the languor, one was struck by the very real effort, made by both sides, to provide a responsible framework within which to judge a major question of public health. It looked to be a fine example of government engaged in the process of looking after the health of its citizens. Consequently, at the end of the long session, one was more than a little disturbed to hear Dr. Überla read a carefully prepared statement as to the BGA's preliminary verdict: this had clearly been prepared in advance of the hearing! The BGA had decided that no rational grounds existed for removing the newer long-acting nonsteroidals from its registry. The drugs were generally safe and effective as anti-inflammatory agents. In due course the manufacturers would be directed to include in their package inserts cautions as to the use of nonsteroidals in the presence of potent cardiovascular drugs. Dosage regimens might be adjusted and the list of clinical indications might be narrowed. The verdict, as enunciated from the prewritten script, seemed unobjectionable to any but the most biased observer. An optimist might have concluded that the BGA had followed Überla's prescription: practical science and practical philosophy had joined in the solution to a "human problem" which resulted from a "breakthrough discovery." But this small exercise of reason seemed to draw the fury of the irrational, and that fury spilled into print.

At the close of the hearing, several of the participants repaired to a central hotel, where an interview had been scheduled with a controversial medical journalist, whom we will call Dieter Gauss. Dr. Gauss's considerable reputation and influence are based partly upon his widely distributed newsletter in which he details perceptions of fraud in clinic and pharmacy. They are also based upon his broad assaults on the BGA in the columns of Germany's most popular newsweekly. In temperament, his writings echo the Luddite convictions of the Greens, a newly powerful political alliance of environmentalists, radicals, and Yankee-baiters. Some foreign observers have detected echoes of brown-shirted National Socialism in their harsh rhetoric, and have waggishly suggested their adherents resemble squares of lawn grass: Green on top, brown on the bottom. We awaited his arrival over well-earned steins of beer.

He arrived, fifteen or so minutes late, introduced himself, and plunked a Japanese tape recorder among the drinks on the table. He was a striking figure. I should have guessed him to be forty-five, but his red cheeks and crew-cut hair made him appear thirty-five: he dressed like a twenty-five-year-old. He wore a tatty leather jacket over a raveled sweater which topped workman's corduroys. Pugnacious in aspect, voice hoarse, alert as a mongoose, he reminded one of saucy Bertolt Brecht:

> *Dabei wissen wir ja:*
> *Auch der Hass gegen die Niedrigkeit*
> *Verzerrt die Züge.*
> *Auch der Zorn über das Unrecht*
> *Macht die Stimme heiser. Ach, wir*
> *Die wir den Boden bereiten wollten für Freundlichkeit*
> *Konnten selber nicht freundlich sein.*

> (Therefore we knew well:
> Even hatred of underprivilege
> Distorts the features.
> Even anger at injustice
> Makes the voice harsh. Alas, we
> Who wanted to lay the foundations of friendship
> Could not ourselves be friends.)
> —from "To Posterity

In that elegant hotel bar, filled with international businessmen in double-breasted suits, it was not an altogether unattractive posture. He looked intensely about the table. Recognizing two executives of an American-based pharmaceutical firm, he growled:

"Why do you sell r-r-at poison?"

The executive replied that they had agreed to an interview, but had no way of handling questions like this.

"No," he insisted. "I am a journalist. I must have answers to my questions. Why do you sell r-r-at poison?" In gentle riposte, one of us asked him whether he was a doctor.

"Ja! I have practiced several years. I was a real practicing doctor! Now I wish to know why you sell r-r-at poison?"

The likelihood was pointed out to him that more humans had been helped by rat poison (the anticoagulant coumadin) than had been harmed. It was also suggested that similar interviews with responsible journalists in English-speaking countries were usually carried out under different conditions of discourse.

"But here we are in Berlin. And I only want facts. You are selling poisons, *nicht wahr?* I mean we have these deaths from your drugs. We heard of death today, from which you make money?"

I replied that some of his interviewees had no connection with the sale of drugs, and that if he had listened to the hearings today it was made clear that fatalities were extremely rare. Administration of any drug, even one as widely used as aspirin, carries *some* risk: it was the job of a doctor to decide with his patient whether the benefits of a medicine warranted that risk. Curiously, this tribune of the people snapped back:

"No. It is the doctor, the drug company who is responsible. The patient cannot know. He is uneducated. You are prescribing poison. You are selling poison."

Well then, we suggested, surely he must have heard Dr. Brune. Even if he discounted testimony from English and American doctors, and from industry, how had he responded to Brune's remarks?

"Ach! He is not a *Facharzt.* He does not even treat patients. *Er ist ein Kaninchendoktor* [a rabbit doctor]. I pay attention only to those who see patients."

I got angry at this one. Gauss was beginning to look less like Brecht and more like Maximilian Schell in *The Young Lions:* a regular fanatic. I blew my cool:

"Dr. Gauss, I should have thought that one of the historical problems of German medicine is that for several years you seem to have performed more experiments on humans than on *Kaninchen*. Perhaps you might wish to think more about animal experiments. For example," I said, pulling out my slide, "There is this remarkable tale of the Australian . . ."

"Ah! You see," he interrupted, "that is all not important. What is important here is the facts. The fact is that your drugs are killing people, and this hearing is only the beginning of a long story. I am very powerful. I was responsible for taking drug X from the market, and I have much influence over what is prescribed. So you should answer my questions. These hearings, here in Berlin today, are only the beginning. Soon they will have such hearings in Paris, in London . . ."

My turn to interrupt: ". . . *und Morgen die ganze Welt!* (and tomorrow the world!) We have heard that one before."

In retrospect it is difficult to decide whether I was angrier at Gauss's echo of the old motto or at his refusal to let me tell him about the frogs. Perhaps Überla and Gauss were telling me something: There was no way to bring the instructive parable of the frogs to the attention of either. There was little left to discuss; the slide went back into my pocket. He returned to the executives.

"But why do you sell r-r-at poison?"

At this point it had become clear to the interviewees that no useful purpose could be served by continuing the session. Whatever model of the journalistic enterprise had settled in the conscience of Dr. Gauss, it clearly bore no relationship to any of our previous experiences with the press.

We parted from Gauss and his tape recorder without the usual courtesies. As he sat there he looked less like Maximilian Schell and more like one of the noncoms in *Stalag 17*. It took a very good dinner, indeed, and a good deal of beer to wash away that long day.

In the weeks to come, Gauss was to pursue his battles with the BGA, printing his unique view of the hearings. His newsletter on the subject makes the *National Enquirer* look like *The Proceedings of the Modern Language Association*. But while he was putting his fancies to paper I was taking off from Tegel Airport. The day was a clear one and we could see Berlin below. The Wall was far less imposing from the air; it seemed an impermanent intrusion on the urban map. In parts, from our vantage, it was hidden from view by factory smoke and the last trees of autumn. In the city divided by that Wall, I had seen East German VoPo's patrolling their zone with automatic rifles, and watched truncheon-bearing West German police monitor crowds of angry protesters against American missiles. At times, the city seemed less an outpost of Western democracy than a memorial to political misfortune. Nevertheless, although haunted for forty years by ghosts of the two major socialist heresies, National Socialist and Leninist, Berlin has managed to settle into the difficult patterns of a free society. After the sad decades of polarized discourse, of Rosa Luxemburg and Joseph Goebbels, Bertolt Brecht and Leni Riefenstahl, reds and browns, the controversies between Überla and Gauss are almost refreshing. Only the rhetoric of the journalist echoes voices of the past. Indeed, judged by this short encounter with the machinery of government, things looked hopeful for the party of reason. And Berlin, divided by Gate and Wall, seems to have provided the lesson for its new democracy that unreason, too, can raise monuments to its triumphs.

When we reached cruising altitude, I picked up Tyler's reprint again. Perhaps, I thought, reverting to the experimental musings of my profession, we could ask the Australians to send us some of these frogs so that we might test whether nonsteroidals would inhibit gastric brooding. But a passage in the article put a fast stop to these thoughts:

> Because the species *(R. silus)* is near to extinction, our data in this and previous reports are based on studies of only five female frogs collected over 8 years.

The order of nature appeared to be as indifferent as the burghers of Berlin to the experiments that amuse one the most. But perhaps there are small families of frogs that have remained undiscovered, hiding under lily pads Down Under, giving oral birth to their young in deep, amphibian satisfaction. If not, it will be necessary to invent them.

Nobel Week 1982

THE ceremonies that crown the Nobel Prize awards in Stockholm resemble a mixture of the Academy Awards in Hollywood and a coronation. Although the names of the recipients are announced in October, the prizes are not awarded until the anniversary of the death of Alfred Nobel, December 10. A brief, handwritten will left the bulk of Nobel's estate to endow prizes in physics, chemistry, physiology, medicine, literature, and peace. The prize for peace is awarded in Oslo, also on December 10, but all of the others are given out in the Swedish capital. In the eighty-one years since the first prizes were given, the Nobel Festival (as it is referred to in Sweden) has assumed aspects of a national celebration, an international jamboree of the intellect, and a remarkable spectacle, at the core of which stands reward of human merit. It is also a great house party.

A guest, arriving early on Wednesday before the major ceremony of Friday night, is likely to be whisked to the Grand Hotel by one of the fleet of black limousines assigned to each of the laureates for a week. The grande dame of Swedish hostelry has been entirely booked by the Nobel Committee, and the front entrance is banked by the large, black Mercedes limousines, each with its identifying placard on the windshield: NOBEL-BIL. The lobby is awash in an untidy flood of radio and television newsmen. This year, the battalions are swollen by scores of Colombians; their ranking author, Gabriel García Márquez, has won the Prize in Literature. In their scurry with lenses, in their scuttle with sight lines and cameras, in their show-biz attire, which contrasts so greatly with that of the other Nobel guests, these Latin reporters appropriately remind one of Márquez's gypsies in *One Hundred Years of Solitude*.

By now it is midday. Barely time to unpack and change. Across the water from the Grand, one can see the Royal Palace, lights already lit, topped by blue and yellow national standards aflutter against the gray sky. The weather is surprisingly temperate for a Scandinavian December; the first snow has yet to fall.

By half past two, the guests of the Laureates in Physiology or Medicine are taken to the Karolinska Institutet, where these Nobel lectures will be given. The lectures in chemistry, physics, and economics are given simultaneously before the approprate faculties; the speech of Márquez will be given at the Swedish Academy that evening. By two-forty-five, the main lecture hall of the Karolinska has been packed for half an hour. Hundreds of students, postdoctorates, and junior faculty fill the seats and spill into the aisles. The front rows, reserved, now fill rapidly with laureates, their families, guests, and the professorial cadre of the Karolinska: Von Euler, Pernow, Luft, Uvnäs, Holm, Böttinger, Ernster, and others. The audience examines itself with much craning of necks; signs of recognition and welcome are given. For the front rows are also studded with visitors from overseas who have been invited by virtue of their close association with prostaglandin research or with the laureates themselves: Flower and Moncada from England; Ferreira from Brazil; Griglewski from Poland; Goodman, Weisblat, and Oates from the United States; Paoletti and the Folcos from Italy. The faces are familiar from a dozen tribal assemblies of the prostaglandin clan.

Silence. The professor of biochemistry extends his greetings and lays out the ground rules. Each of the laureates will give a lecture of 30 to 40 minutes, followed by a ten-minute intermission. At the close of the third lecture, there will be a gala reception. (A glance at the audience assures the visitor that all are suitably dressed for standing about with champagne.) The professor introduces Sune Bergström, who is being honored for bringing prostaglandins—fatty substances which function as local hormones—from the laboratory to the clinic. Senior of the three prizewinners, the magisterial Bergström outlines with great modesty his achievement in isolating prostaglandins and defining their chemical structure. As befits a major official of the World Health Organization, he describes how prostanoids can modulate fertility and how they show promise in meeting the needs of underdeveloped countries. He addresses a fitting tribute to his former teacher, Ulf Von Euler (Nobel Prize, 1970), now seated before him in the center of the front row. It was Von Euler who gave prostaglandins their name and directed the early work of Bergström. The chain of learning and scholarship is evident.

The tradition becomes yet more palpable in the room as

Bengt Samuelsson assumes the podium after the intermission. For Samuelsson has not only followed *his* teacher, Bergström, in prostaglandin research but succeeded him as dean of the medical faculty. Tall, blond, and soft-spoken, Samuelsson describes the biological insight and bravura chemical analyses that, in addition to elucidating the biogenesis of prostaglandins, led to the discovery of related and equally important compounds, thromboxanes, and leukotrienes. He finishes with a dazzling summary of how these products play vital roles in the pathophysiology of circulatory disorders, inflammation, and asthma.

John Vane, of England, follows the two Swedes. With great wit and commensurate affability, the portly Vane generously splices vintage photos of his collaborators into the narrative of his main discoveries. By means of ingenious *in vitro* techniques based on those described by *his* intellectual mentor, Sir Henry Dale (Nobel Prize, 1936), Vane found that another substance related to prostaglandins, prostacyclin, was a key component in the regulation of vascular tone and the clotting of blood. He also made the heuristically important observation that aspirin and similar drugs inhibit the biosynthesis of prostaglandins. Vane finishes his lecture with a reassuring message: In an era of intricate instrumentation and recondite stereochemistry, the methods of classical pharmacology can still yield great results in the service of a prepared mind. Applause. The lectures are history.

The auditors are now shepherded through the halls of the Institutet by students dressed in folk costume. Other groups are singing. Outside, it has been dark since three-thirty, and small pots of flame outline the driveways. Inside, large candles are lit, as the passageways yield to a large, blond-wood reception hall. Hugs, kisses, or handshakes are exchanged on the receiving line formed by the laureates and their families. And then the corks pop and champagne flows. It appears that all the cellars of Rheims have been abruptly emptied for this, as indeed for all the week's receptions. As the Vikings of the Karolinska fall upon the ever-filled tulip glasses, the animated crowd mixes with little regard for the usual hierarchies of Swedish academia. Postdocs chat with laureates, teenagers with distinguished visitors, laboratory assistants mingle with professors.

The reception also serves as an introduction to the bounty of Swedish hors d'oeuvres: dried or marinated slivers of reindeer meat, coral beads of salmon caviar on sliced eggs, a most imaginative assortment of herring, and butter-soft smoked salmon. In multiple geometric disguises these appetizers will reappear in the course of receptions and dinners to follow this one. Indeed, the taste of champagne and reindeer meat will become as engraved on the gustatory memories of the Nobel guests as *madeleines* on Proust's. As the champagne flows and music plays, it is difficult to remember that the way to this brightly lit hall has been by means of long hours at the bench, the disappointments of false leads, and more than one hundred man-years of solitude.

By half past seven, the reception gradually breaks up, and the jet-lagged guests are dispersed to various dinners in the city. Limousines drop the elated groups into charming eateries of the Old Town snuggled in cellars behind the Royal Palace. There, too, reindeer and salmon, white wine and aquavit extend the evening.

Meanwhile, as subsequently reported in the papers, Gabriel García Márquez is addressing a packed house in the

Swedish Academy, which flanks the palace. The fiery, mustachioed, and close-cropped Colombian delivers a speech charged with the energy of the Latin American left. He is reponsible for the presence of Régis Debray and Mme. Mitterrand at the festival. They are in the audience as the anti-Yankee writer pays homage to his literary, if not ideological, kinsman:

> On a day like today, my master, William Faulkner, said, "I decline to accept the end of man." I would feel unworthy of standing in this place that was his, if I were not fully aware that the colossal tragedy he refused to recognize 32 years ago is now for the first time since the beginning of humanity nothing more than a simple scientific possibility.
>
> Faced with this awesome reality . . . we, the inventor of tales, who will believe anything, feel entitled to believe that it is not yet too late to engage in the creation of the opposite utopia, a new and sweeping utopia of life, where no one will be able to decide for others how they die, where love will prove true and happiness be possible and where the races condemned to 100 years of solitude will have at last and forever, a second opportunity on earth.

A few houses away, the evening is ending for the visitors from overseas, as buoyed by adrenalin and Veuve Cliquot, they toast the science of medicine. The first night ends; sleep is more than welcome.

On the next day, the first scheduled activity is a reception at the Nobel Library of the Swedish Academy at 3:30 P.M. This information is found on a personalized schedule, which is clipped to a sheaf of engraved invitations, admission tickets, and arrangements for transportation. And so, after a morning of museum hopping, the overseas visitors are brought a few blocks to the Swedish Academy. By mid-afternoon the sky is already darkening; a gentle rain falls. Up the stairs and past a receiving line, at which towering officials of the Academy and the Nobel Foundation offer courteous greetings, one is ushered into the Nobel Library. The room is blue and white—it is pure eighteenth century and large enough to accommodate 200 to 300 guests.

This is the first occasion at which the entire group of visiting guests and laureates has assembled. One can identify among them Ken Wilson of Cornell, the Prize winner in Physics, and the large Wilson clan. Wilson looks a bit like a solid Warren Beatty, and he is surrounded by a crowd of dazzled well-wishers. But so is his father, E. Bright Wilson, who is one of America's best-known physical chemists. And there is scholarly Aaron Klug, the Cambridge Prize winner in Chemistry, chatting in animated fashion with one of the sequined ladies of South America. Tall, bankerish George Stigler, the Laureate in Economics from the University of Chicago, chats with Daphne Vane, whose husband is busy having his hand pressed by a Nobel official. On this occasion, dress is not formal, but splendid cocktail gowns are worn by the women and sober, dark suits by the men. The ubiquitous reindeer, salmon, and champagne are dispensed in the main chamber, where under a glass case lies the original will that started the Foundation. A bust of Nobel is at the end of the room, and the guests are assured that this likeness—as well as all the other portraits and busts that dominate the interior of the Nobel Foundation—were executed from photos, since Nobel never consented to pose for painter or sculptor.

Groups of guests wander into the Nobel Library itself, where the works of Yeats nestle against those of Steinbeck, where Darwin and Dickens share an alphabetical fate. But the mix of living guests is as random as the world of learning and scholarship: Economists lift glasses with physicists; a few biochemists and some of the Colombian exiles have found a common language in French. Yesterday's reception was a tribute to the enterprise of biomedical research; today we are here to honor the general world of scholarship and literature. No speeches, no formal arrangements, simply an amiable crowd with good heads and no glitz gathered in one of the most exquisite halls this side of *Der Rosenkavalier.*

By 5:00 P.M. the reception is over, and the guests repair to a series of private parties and dinners given by the laureates or their friends in Stockholm. After the large grouping of the afternoon, the scene shifts to traditional Scandinavian hospitality. For some this proves quite grand, indeed. Thus, García Márquez dines as the guest of Olof Palme, the socialist prime minister of Sweden, with Régis Debray and Danielle Mitterrand. Also invited is a former prime minister of Turkey, Bulent Eçevit, who has been paroled by his military junta for the occasion. In contrast, other guests attend more modest receptions at the homes of the local laureates, warm gatherings of old friends and scientific collaborators. The receptions are followed by long, relaxed dinners, which end somewhat early, for tomorrow is the big day—the day of the ceremonies.

The Grand Hotel bustles with activity on the next morning. It is Friday the tenth of December. The women's hairdresser has been booked for weeks, and the scene is said to resemble the dressing room of a grand opera three minutes before the curtain is raised for *Aïda.* An American guest, to whom the staff has allotted a precious few minutes, stares with wonder as a distinguished lady, addressed as Mme. Nobel, has her coiffure studded with networks of real diamonds. The men's costumes are also being prepared. Hotel valets swish about the corridors carrying newly pressed tailcoats, white pique vests, and satin sashes as if the safety of a country depended upon their timely and unruffled delivery. In the coffee shop, Aaron Klug prepares for the evening-long ceremonies with a quick hamburger. Everyone must be ready by four in the afternoon, for the ceremonies will begin at four-thirty sharp.

The *Nobelstifelsens Hogtidsdag* (or Solemn Festival of the Nobel Foundation) is held in the Grand Auditorium of the Concert Hall. Many thousands wish to attend, but there are only 1,070 seats in the hall. The laureates are each assigned twenty-five for family and guests.

Rain is pouring at four. The lobby of the hotel is filled with guests in gala attire waiting to be transported by car or bus to the ceremonies. The pearls and diamonds, the medals and white ties are covered by cloak or fur. The busy press and television photographers light the scene with flash and spotlights. Everyone is then trundled off to the Concert Hall.

Despite the rain, a crowd of onlookers has already gathered to view the arrival of the glittering crowd. And in the lobby, a transformation is achieved. As the cloaks and capes are surrendered to wardrobe attendants, the guests are revealed as characters out of Franz Lehar: We have entered the past century. The women wear lavish ball gowns of silk and taffeta. The hairdresser's hand is evident in the ribbons and flowers braided into blond Swedish hair and in the gems that gleam from the high, white coiffures of the academic dames. There is no question, however, but that it is the men who steal the show. The printed directions for dress mandate white tie and tails. And although the Americans feel quite splendid in their rented black-and-white costumes, they are positively dowdy compared with the Europeans or South Americans. For these worthies have arrived swathed in medals and decorations: red and blue sashes worn diagonally over white vests, scarlet ribbons at the neck, enameled gold orders in the shapes of stars, crosses, and sunbursts worn over the breast or impaling a ceremonial shawl. Members of the diplomatic corps glitter like Mexican shrines; the Swedish professors doff the distinctive, academic top hats which bear not a little likeness to those of the Puritan fathers.

The scene is a gala bedlam of glitter and display; one finds it difficult to leave this scene of ritual costuming. After tiaras and sashes are adjusted, after another round of voyeuristic excess, each couple ascends a grand staircase to the main hall. The ushers are all students who, in addition to their formal dress, sport the marshal's red sash and a white student cap. The great hall fills quickly, for the ceremony begins sharply at 4:30 P.M. The stage is brightly lit for worldwide television. It is decorated with masses of flowers from San Remo, the winter refuge of Nobel. His bust is floodlit in the background. In a good-sized gallery above the stage is most of the Stockholm Philharmonic with their conductor, Sixten Ehrling. On twin banks of tiered stage seats are placed members of the Nobel Foundation and an assortment of dignitaries.

The hall is filled and the doors are now shut; the audience is hushed. Again there is a good bit of neck craning and delicate finger pointing but this is interrupted by a sound of trumpet and organ music. The royal family arrives amidst a triumphal fanfare. More trumpets, and the laureates troop in. The King and Queen, Carl Gustaf XVI and Sylvia, seem to have been brought in for the evening from either Central Casting or a fairy tale. He is tall, dark, and handsome. She is beautiful and dressed in blue and silver; ransoms of diamonds are scattered on silk and crown her hair. They are seated, stage left, on two blue chairs, which stand slightly higher than those of two other members of their party. Seated stage right in individual armchairs are the laureates, all except one dressed in white tie and tails. Márquez has chosen the permissible option of "national dress" and wears a tunic-topped, white linen costume, called a *liqui-liqui*. With his cropped hair, mustache, and tunic, he looks like the young Stalin.

More music, and the deputy chairman of the board of the Foundation rises for the first speech. He gives a stirring defense of the importance and influence of the Nobel Prizes in the world of science. He dismisses—without direct reference—a crotchety editorial in *The New York Times,* which grumbled about the winners, the donors, and the extravagance of the Nobel Prizes. Dr. Browalth replied (the remarks are in Swedish, but the audience is given a booklet with English translation): "Criticism of Alfred Nobel's donation in a recently published commentary on this year's prizes has therefore been forced to resort to the method of rewriting history, twisting and misinterpreting facts to make them fit the slanted argument. According to this opinion, the prizes are tainted because Nobel made his fortune as a maker of arms and ammunition. This is not correct. The property left by him comes from his inventions in the field of explosives, which have enabled tunnels to be driven through mountains, facilitated the construction of roads and railways, and laid the foundation of the modern mining industry." This speech is followed by Elgar's *Pomp and Circumstance,* during the course of which the audience has time to ponder the relationship between railways and artillery shells, between science and industry, between prizes and motivation, and of the distance between the Concert Hall and the place where dinner will be served.

The next speaker is Professor Lundquist of the Royal Academy of Science, who briefly summarizes the work of Kenneth G. Wilson. He then proclaims the citation: "For his theory for critical phenomena in connection with phase transitions." The laureate rises, advances to stage center, and receives his Prize in Physics (medal and scroll) from the King, who accompanies the award with a handshake. The King sits, and the laureate—alone on the stage—bows directly to the audience. No tedious words of gratitude and acknowledgment are uttered. A volley of prolonged applause. The ceremony is repeated for Aaron Klug, who receives from Bo Malmström his Citation in Chemistry: "For his development of crystallographic electron microscopy and his structural elucidation of biologically important nucleic acid-protein complexes." More music. Then Bengt Pernow outlines the work of Bergström, Samuelsson, and Vane, citing their "discoveries concerning prostaglandins and related biologically active substances." They receive their awards, one after the other, from the King and bow to massive applause. More music, Bartók this time. Literature is next: Lars Gyllesten, honoring Márquez's apparent desire to avoid the language of the Yankee in this ceremony, speaks in French to Márquez (the remarks are translated into English and Spanish) in citing the author for "his novels and short stories, in which the fantastic and the realistic are combined in a richly composed world of imagination, reflecting a continent's life and conflicts." Márquez receives his prize and, after his bow, gives a great grin, waves to the cheering audience of his adherents, clasps his hands above his head in the manner of a champion bantam weight, and resumes his seat. The music now played is by an American: William Schuman's "Chester" from his *New England Triptych*. A statement?

The last recipient is George Stigler, cited in economics for "his seminal studies of industrial structures, functioning of markets, and causes and effects of publication regulation." A bow, applause, and now the whole assembly rises as the Swedish National Anthem is played. More music; the King and Queen withdraw. The laureates remain on stage, and as the audience begins to leave, the immediate families of the prize winners rush onto the stage to exchange embraces and kisses.

By limousine and bus, the guests are brought to the great Town Hall of Stockholm. Despite the rain, torchbearers light the road to the courtyard. They are Boy Scouts, and their wet faces shine with greeting. After entering from a cloistered *porte cochére,* the guests repeat the disrobing ritual. They are immediately given an elaborate booklet, in which each guest's name, alphabetically listed, has been assigned a number. A detailed foldout seating plan locates his assigned place at one of the tables. The banquet hall in which the *middag,* or dinner, is served is enormous: It is the height of the Metropolitan Opera in New York but larger. The style is red brick and vaguely Florentine (the Palazzo Vecchio comes to mind). Indeed, this banquet hall is still called the Blue Hall after its intended stucco color, but the architect was so impressed with the raw power of uncoated brick that it has remained *brut* ever since.

In the center of this puissant hall stands the long "A" table, set for eighty, at which will sit the laureates, the royal party, the prime minister, and various dignitaries. The other Nobel guests will be accommodated at two flanking sets of twelve tables holding thirty places each. Around this middle grouping are more than forty tables, at which anywhere from eight to forty students are placed. Their presence has traditionally been a feature of the Nobel banquet, and by this means the students are initiated into the splendors and rituals of Swedish academia.

The audience, with glitter and medals, wet hair and high spirits, find their assigned seats. One is not seated with one's spouse or even necessarily at the same table. Instructions are that both the initial conversation and the first dance request are addressed to the woman on the right. The champagne glasses are already filled, the tables are candle-lit, the hall itself is lit by lanterns and reflections from day-bright television spotlights. More trumpets sound, as the royal party and the laureates, arms linked with dinner partners, descend the grand staircase from the balcony above.

After the party at the main table has been seated, toasts are proposed, first to the laureates, then to Alfred Nobel: The hall resounds with *Skoal*! General conversation swells the room, much of it directed toward the seating arrangements. Sune Bergström is next to Mme. Mitterrand, Daphne Vane is between Aaron Klug and the prime minister, Mrs. Wilson is next to the King, as is Mrs. Klug. Bengt Samuelsson is beside Señora Barcha de García Márquez. The Queen is next to John Vane; she is the target of a hundred flashbulbs. Karin Samuelsson is flanked by Ken Wilson and García Márquez. The table is filled with ambassadorial couples from the United States, Britain, Colombia, and Norway.

The guest tables are dotted with names of Sweden's great families: Nobel, Bonnier, Hammarskjöld, and so forth. The Karolinska is well represented, as are *De Aderton*—The Eighteen—who are the Swedish equivalent of the French "immortals" in scholarship and letters. There are tables for the press and dozens of Stiglers, Wilsons, Vanes, Klugs, and Samuelssons. The glasses are filled again.

Service is directed by a kind of regimental maître d', who with a wave of hands orchestrates a battalion of waiters and waitresses. The menu begins with marinated reindeer meat and progresses to a species of salmonlike trout, fresh from the Arctic, over dill and rice. The only wine is champagne: Rheims is surely dry tonight. The lights dim as a vast procession of servants descends the great staircase. Above their heads they bear the famous Nobel parfait—an elaborate concoction of two kinds of ice cream and spun sugar, each portion of which is topped by a candied *N*. Port is passed with dessert, and conversation rises again.

A sound of trumpets. Silence again. It is now the turn of each laureate to say a few words; the speeches are no longer than three minutes. In English, Bergström speaks of science, Vane of discovery, and Stigler of the art of economics; García Márquez speaks in Spanish of the power of poetry. Ken Wilson seizes this moment, as the United States Nobel Laureate in Physics, to call for an end to the development of nuclear technology for war. And Bengt Samuelsson brings us back to the academic basis of the pomp and glitter. He addresses the students and welcomes them to a fellowship in which scholarship and merit are their own rewards, to a fellowship of peaceful study, and to the possibility of high achievement. Applause.

A sound of trumpets. Silence, a crescendo of music, and the stairway is awash in light. An amazing procession now descends. It is led by a flag-bearing platoon of Swedish students in red sashes, tailcoats, and their distinctive white caps, but behind them comes a never-ending array of Latin American dancers, singers, and entertainers. In this tradition-breaking display, the government of Colombia, perhaps to entice García Márquez back from exile, has flown to Sweden several troupes of popular artists. These now proceed to perform what can only be described as a wild, rhythmic night club show of music, song, and dance. Drums beat in the northern hall. Colombia may have suffered one hundred years of solitude, but it cannot have endured these in silence! Tepid applause rewards these efforts.

After this unusual interlude, the customary entertainment proceeds. This is provided by a chorus of Swedish students, who sing Swedish folk songs, American spirituals, and academic anthems. Loud applause, the house lights are turned on, and the dinner is over.

It is time for the dancing to begin. The guests now ascend the great staircase that leads to the large *Gyllene Salen*—the Golden Salon. This marvelous hall is entirely faced with golden mosaic tiles, the decoration of which, by means of allegorical flora and fauna, follows a style that successfully blends Gustav Klimt with Byzantine Ravenna. The orchestra begins with great swoops of waltzes. The floor is soon filled with billowing skirts and a flapping of tailcoats. The bars are open, the students join in the dance, and the music moves from fox trot to swing to disco as the evening unfolds.

The dances continue long after midnight; no one wants the evening to end, and guests reluctantly straggle to the waiting cars. Many roam the salons of the Grand Hotel in search of the final brandy at 2:00 A.M.

There is more on tap for the weekend. The laureates will receive their checks—$150,000 per prize—in a flutter of signatures and notaries. Later, on Saturday night, they will dine alone with the King and Queen at the Royal Palace, attended by liveried footmen. There will be small, private farewells for their overseas guests at the Grand Hotel; García Márquez will entertain his well-wishers to Latin rhythms. There will be performances by the Royal Ballet at the jewel box of an opera house and long leisurely holiday lunches at the *Operakällaren*. But one other event will touch the memories of many.

For the past decade or so, it has become customary for the Jewish community of Stockholm—some 9,000 or so—to dedicate its Saturday service to the laureates who are Jewish. And on this Saturday, which happens to be the first day of Chanukah, two of the laureates and their guests assemble at the two-hundred-year-old synagogue, a few blocks from the Grand Hotel. Outside, as before many such buildings in Europe, a precautionary police van is parked. Inside, the first candle has been lit.

This is a traditional congregation, with the women seated upstairs. As the service unrolls, a small boy who has just celebrated his thirteenth birthday reads the Torah. The rabbi advises him that one model of the good life is that exemplified by Aaron Klug, who has preceded the boy in read-

ing the book. At the end of the ritual, the scrolls are returned to the Ark, and as the curtains are unfurled, it comes
as a kind of revelation that it is the Book, the Word, that
are the subject of worship in this place. As learning and
scholarship are honored in the secular rituals of Friday, so
the Word is honored on Saturday in this northernmost
outpost of the Diaspora. The rabbi, an American from
Philadelphia, gives a stunning sermon for the occasion. It
is in English and is laced with apt quotes from Socrates
and Gershon Cohen. He declares that the feast of light, the
enterprises of arts and science as honored by the Nobel
Prizes, and indeed the Sabbath itself are manifestations of
God's gift of the Word: the gift of human reason. For the
auditors—Swedish, American, English—some of whom have
not been in touch with this religion since they themselves
were thirteen, this celebration of reason seems a fitting end
to the winter ceremonies.

Against *Aequanimitas*

THE week before our graduation from medical school in the midfifties, one of the drug houses gave each of us a book of essays by Sir William Osler, entitled *Aequanimitas*. For most of us, this was the first book of medical essays we had encountered, and I remember it as pretty stiff going indeed. Attracted to medicine in no small part by the sentimental heroics of *Arrowsmith, Microbe Hunters* and *The Citadel,* we were not about to swallow Osler's call to stern duty. *Aequanimitas* and the other essays are earnest, humorless, and strict. Written in the magisterial prose of the classicist, they are laced with highfalutin' references to the orators of Rome and the pantheon of Greece. They are by no means without interest today.

Rereading these dated classics of medical discourse, it is not difficult to understand why the figure of Osler has dominated the traditions of our profession for so long, why his shade lurks in the libraries of our glittering institutions, or why his name still evokes a certain glory. Osler's reputation was forged in three countries, in four schools, and by hundreds of well-placed students. Successively, Professor of Medicine at McGill, Pennsylvania, Johns Hopkins, and Oxford, Osler presided over the emergence of Anglo-Saxon medicine as an empirical discipline which rivaled the clinical imagination of the French and the scholarly rigor of the Germans. From the 1880s to 1920, Osler and his school transferred American medical teaching from the lecture theater and private clinic to the supervised bedside of university hospitals. Osler and his colleagues at Hopkins created the American medical school as we know it today.

Osler also wrote an enormously successful textbook of medicine which set the intellectual standards of the day; he made significant contributions to the clinical analysis of bacterial endocarditis, angina pectoris, vascular diseases, and the finer points of physical examination; taught daily; ran popular ward rounds and clinics; served on innumerable commissions; was consulted on the diseases of presidents, tycoons, and poets—he treated Walt Whitman for dyspepsia—became president of the British Classical Society in 1919; was knighted; and found time to produce scores of belletristic essays. In *The Old Humanities and the New Science* (1920) he displays the energy which fueled his prodigious talent, summarizing an era before the British Classicists:

> To have outgrown age-old theories of man and nature, to have seen west separated from east in the tangled skein of human thought, to have lived in a world remaking—these are among the thrills and triumphs of the Victorian of my generation . . .

What optimism, what faith, to express satisfaction with his generation at the end of that dreadful war into which it had blundered! In that war, the audience he was addressing had witnessed the loss of much of the next generation in the trenches of Flanders: Osler himself had lost a son. Never one to permit personal experience to modify public convention, Osler continued:

> No wonder war has advocates, to plead the historical clash of ideals, the purging of a nation's dross in the fire of suffering and sacrifice, and the welding in one great purpose of a scattered people . . .

The Victorians of Osler's generation were driven by a remarkable devotion to social conflict, to strife and competition, to the clash of man against man, to the "battle of ideas." In short, to social Darwinism. This unpleasant view, which dominated much of late nineteenth-century thought, has been carefully examined by modern cultural historians who explain its attractions to the professional classes. After the hierarchies of social rank had been dismantled by the revolutions of the eighteenth century, the mercantile classes—from which the professions derived—framed new categories by which the worthy could be distinguished from the undeserving. What better model than the genetic? Had evolution not been suggested by the evidence of species, the Oslerians would have had to invent it. Wallace and Darwin may have made *biological* discoveries, but the social implications of those discoveries were very clear to Osler. Osler assures us that:

> In no way has biological science so widened the thoughts of men as in its application to social problems. That

throughout the ages, in the gradual evolution of life, one unceasing purpose runs; that progress comes through unceasing competition, through unceasing selection and rejection; in a word that evolution is the one great law controlling all living things, "the one divine event to which the whole creation moves," this conception has been the great gift of biology to the nineteenth century.

It is remarkable that this sentiment issued from one who was witness to the triumphs of the sanitary revolution, the decline of the great American epidemics of yellow fever, malaria, typhus, cholera—and who popularized the rise of modern bacteriology with the work of Lister, Koch, Pasteur, and Wright. The perception by Osler that it was the *social* lesson of evolution which was the greatest gift of biology to his century is a telling example of how even the keenest among us tends to see the world through lenses ground by the temper of his time.

Osler was not only familiar with the classics of Greece and Rome, but read current biological and medical literature with quick insight. I am amazed that within a few years after Elie Metchnikoff had discovered phagocytosis at the seaside of Messina, Osler was delivering a lecture on the subject at my hospital. He described to students and physicians of Bellevue how phagocytes disposed of bacterial debris, and critically examined the role of white cells in malaria. His admonitory lectures are laced with references to English, German, and French journals of the time; they contain generous appeals to young doctors which urge them to overcome local chauvinism. He prepared reading lists of "great books," insisting that the physician must study world literature as well as the techniques of his craft.

And, yet, when we read the essays of this accomplished

man, the major voice that seems to emerge from all this serious, uplifting advice is the public tone of the academic snob. Osler was perhaps the quintessential club member. Comfortable in the wing chairs of the paneled mansions which sheltered Victorian men from women and the lower orders, he was instrumental in founding at least two clubs of academic physicians that thrive to this day, the Association of American Physicians and the Interurban Clinical Club. Membership in each is still prized as a kind of seal of scholarly approval; happily with time, intellectual merit has replaced good table manners as a requirement for membership. But for Osler, the ideal of the doctor as a patrician, whose *gravitas* was an element of his pharmacopoeia, was simply a projection of the code of the club. At Hopkins and at Oxford, phthisis may have racked the lung, dropsy quenched the heart, and sepsis seized the brain, but the observing physician—who had not one drug capable of meaningful intervention—was required to remain imperturbable. Again, from Osler:

As imperturbability is a bodily endowment, I regret to say that there are those amongst you, who, owing to congenital defects, may never be able to acquire it. Education, however, will do much; and with practice and experience the majority of you may expect to attain to a fair measure. The first essential is to have your nerves well in hand. Even under the most serious circumstances, the physician or surgeon who allows "his outward action to demonstrate the native act and figure of his heart in complement extern," who shows in his face the slightest alteration, expressive anxiety or fear, has not his medullary centers under the highest control, and is liable to *disaster* [my italics] at any moment.

This concern for *gravitas,* for *aequanimitas,* for the Stoic and Epicurean, is a feature of much of the Oslerian legacy. It is a record of his generation, the energy of which hammered out the structure of American medical education. Most of us are unsure whether that structure remains useful, and to our students the concept of *aequanimitas* must appear as quaint as grammar itself. The reason for our rejection of *aequanimitas* as a mode of conduct is our own experience of history in this bloody century. After Guernica, Dachau, Hiroshima, and My Lai, Osler's advice, which urges us to pay greater attention to "disasters" of social appearances than to avoiding failures of conscience, seems less than compelling. Few would now agree that loss of face is worse than loss of feeling. Osler warns us of another danger:

> There is a gradually accumulating surplus of women who will not or who cannot fulfill the highest duties for which Nature has designed them. I do not know at what age one dare call a woman a spinster. I will put it, perhaps rashly, at twenty-five. Now, at that critical period a woman who has not to work for her living, who is without urgent domestic ties, is very apt to become a *dangerous element* [my italics] unless her energies and emotions are diverted in a proper channel . . .

This passage, and Osler's devotion to "unceasing selection and rejection" come to mind when we are faced with the minutes for June 5, 1901, of the Medical Faculty Advisory Board of Johns Hopkins:

> The Dean presented the marks of the Graduating Class and on motion of Dr. Osler it was voted that Miss Gertrude Stein be not recommended for the degree of Doctor in Medicine.

No, the vision of Osler, a vision in which unceasing social struggle can only be relieved by the grandeur of personal equanimity, must strike the reader of today as having
failed the test of history. The defects of that vision were
perhaps best described by E. M. Forster in his essay on "The
English Character":

> And they go forth into a world that is not entirely com-
> posed of public-school men or even of Anglo-Saxons, but
> of men who are as various as the sands of the sea; into a
> world of whose richness and subtlety they have no concep-
> tion. They go forth into it with well-developed bodies, fairly
> developed minds, and undeveloped hearts.

Osler does not entirely warrant this description: Cushing's biography of Osler is generous in his account of Osler's loyalties to colleagues, students, and subordinates. Like
that of most Victorians, Osler's domestic life was not a
subject for public description. Cushing devotes pages to each
tedious engagement of Osler with this assembly or that
ceremonial dinner; only a paragraph or two is devoted to
Osler's marriage day, which, late in life, seems to have been
arranged at a moment of convenience with respect to his
professorial commitments.

But it is neither a lack of personal charity, nor of civic
sentiment, that troubles us when we read Osler today. His
ear, so sensitive to cultures of the classic past, was deaf to
voices from the street or the studios of his own time.
Whitman was incomprehensible to him—too wild, too
unorthodox. The rise of social thought was an intrusion.
Not for Osler the throbbing of new art or new music, the
stirrings of the melting pot. One senses that when he finally reached Oxford from America he found his snug haven.

Out-donning the dons, he celebrated the unchanging life of the academy in a confusing world over which his generation had lost control. Having won an academic revolution in the New World, he spent the remainder of his career writing solemn appreciations of the culture of the Old.

The results of that revolution at Hopkins, secured two generations ago by acceptance of the Flexner report, are under assault in the 1980s. Osler and the men of Hopkins, who helped to transform medical schools from a collection of practical workshops into faculties of research universities, based their educational methods upon the twin foundations of laboratory instruction and direct clinical experience. They assumed that students would have come to their schools with a classical education which had prepared them for graduate study. But in the last decade, faced with sometimes undeniable accusations that our graduates are greedy, overspecialized, unfeeling, and uncultivated, medical schools of the 1980s have responded by elaborating courses in medical ethics, medical sociology, and medical humanities. Time for these has been taken from laboratory and clinical experience: lectures on ethics have replaced hours at the bench. Harvard is about to launch such an experiment for a portion of its class, bravely called "a new pathway."

These programs, in what I would call "remedial humanities," seem to be a sorry substitute for a sound undergraduate education in the liberal arts. Our admission policies and examination procedures have apparently been designed to guarantee the successful candidacies of wonks. It therefore seems unlikely, to me at least, that didactic instruction in remedial ethics or "medical humanities" will turn these successful maze runners into caring physicians.

Moreover, I have serious doubts that tutored appreciation of the arts, of the humanities, or of philosophy will, like monosodium glutamate, enhance the flavor of a student's humanity. Perhaps we should feel uneasy when it is argued that we can change people for the better by exposing them to strong doses of ethics, or humanities, or even molecular biology. Did his well-formed phrases, his quotes from Plutarch, turn Osler into a compassionate crusader or endow him with a social conscience? Did his readings in Ovid transform him into an overt sensualist? Did his excursions into late Roman philosophy prevent him from pointing out to his house officers that there are three sexes, "men, women, and women physicians"?

No, I'm afraid that formal instruction in ethics, aesthetics, or the "larger issues" of the humanities will not assure the goodness—in the general sense—of our students or teachers. Remember Alexis Carrell? A far more distinguished scientist than Osler (Carrell won the Nobel Prize in 1912 for his work in vascular surgery), he was also a prolific essayist and rabid "humanist." His social Darwinism, so much a product of the Oslerian school, extended to such examples, taken from "Man the Unknown":

> Indeed, human beings are equal. But individuals are not. The equality of their rights is an illusion. The feeble-minded and the man of genius should not be equal before the law. The stupid, the unintelligent, those who are dispersed, incapable of attention, of effort have no right to a higher education. It is absurd to give them the same electoral power as the fully developed individuals. Sexes are not equal.

Carrell, at the end of his life, voluntarily returned to work for the Vichy regime at an institute of which the function was the study of applied eugenics. One of his students was the only graduate of the Rockefeller Institute to have been indicted as a Nazi war criminal for human experimentation. Carrell, a cultivated, urbane, classically trained scholar, finally voiced the latent message of social Darwinism. Cool and dispassionate, he spelled out his version of *aequanimitas* from which it is clear that his view of women is superimposable upon that of Osler.

A great race must propagate its best elements. However, in the most highly civilized nations reproduction is decreasing and yields inferior products. Women voluntarily deteriorate through alcohol and tobacco. They subject themselves to dangerous dietary regiments in order to obtain a conventional slenderness of their figure. Besides, they refuse to bear children, such a defection is due to their education, to the progress of feminism, to the growth of short-sighted selfishness.

Note the idealistic appeal to selflessness! Indeed, Carrell, who expounded his views on medical education not only at the Rockefeller Institute, but also to his artsy friends of the Century Association, was sure that good stiff doses of Nietzsche, Spengler, and Darwin would turn the over-technical doctors of his day into well-rounded recruits to the elite armies of eugenics.

I do not mean to ridicule Osler's gentle scholarship by appeals to Carrell's social mischief. What emerges, however, from the works of these capable physician-educators is that neither broad cultural achievement nor intellectual skills alone can protect us from nasty social views or simple obtuseness. No courses in social philosophy, in medical ethics, or medical literature will assure us that our students will become caring, compassionate, or humane. Those values should be part of the matrix of our communal education, introjected by example and learned before the first MCAT is taken. Neither the humanities nor the sciences are value-free, and both should disturb as well as comfort. Both, when splendid, should make us less comfortable with what we find in ourselves. But if the goal of the new medical humanities is that of the old, if its aim is to lead to *aequanimitas*—to learn control over our "medullary centers"—I want no part of it. The passion of the physician may be the best part of what he has to offer his patients and his society. When we repress the urging of our passions in the hope of realizing some vague ideal of the physician, we risk tripping the Oslerian booby trap. Dylan Thomas may have gotten it right: rage, not *aequanimitas,* ought to be the proper human response to disease and death. Finally, the Oslerian view is not only devoid of passion, but of joy. I wonder whether those futile, remedial courses of medical humanities, about to be launched in Boston, will include this antidote to *aequanimitas* written by Dr. William Carlos Williams, who entered the University of Pennsylvania School of Medicine just two years after Osler left for Hopkins:

> *If I when my wife is sleeping*
> *and the baby and Kathleen*
> *are sleeping*
> *and the sun is a flame-white disc*
> *in silken mists*
> *above shining trees,—*
> *if I in my north room*
> *dance naked, grotesquely*

THE WOODS HOLE CANTATA

before my mirror
waving my shirt round my head
and singing softly to myself:
"I am lonely, lonely.
I was born to be lonely,
I am best so!"
If I admire my arms, my face
my shoulders, flanks, buttocks
against the yellow drawn shades,—
Who shall say I am not
the happy genius of my household?
          —from "Danse Russe"

Source Notes

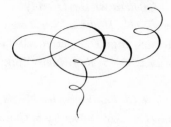

The Woods Hole Cantata

Duffus, R. L. "Jacques Loeb." *Century Magazine,* Vol. 2, pp. 374–385, 1925.

Loeb, J. *The Organism as a Whole: From a Physico-Chemical Viewpoint.* New York: G. Putnam's Sons, 1916.

Loeb, J. *The Mechanistic Conception of Life,* reprint ed., Cambridge: Belknap Press of Harvard University, 1965.

Loeb, J. "Biology and War." *Science,* 45:73–76, 1917.

Osler, Wm. *The Old Humanities and the New Science,* intr. H. Cushing. Boston & New York: Houghton Mifflin, 1920.

Osterhout, W.J.V. "Jacques Loeb," *Journal of General Physiology,* 8:9–42, 1928.

Foucault and the Bag Lady

Foucault, M. *Madness and Civilization.* New York: Random House, 1965.

Foucault, M. *Birth of the Clinic: an Archaeology of Medical Perception.* New York: Pantheon, 1973.

Laing, R. D. *The Divided Self: A Study of Sanity and Madness*. London: Tavistock Publishers, 1960.

Pinel, P. *A Treatise on Insanity*, 1806, trans. D. D. Davis. New York: Hafner Publishing Company, 1962.

Cholera at the Harvey

Duffy, J. *A History of Public Health in New York City 1866–1966*. New York: Russell Sage Foundation, 1974.

Giono, J. *The Horseman on the Roof*. New York: Alfred Knopf, 1953.

Rosenberg, C. E. *The Cholera Years; the United States in 1832, 1849 and 1866*. Chicago: University of Chicago Press, 1962.

Vaughan, M. "Choleragen, adenylate cyclase and ADP-ribosylation." *Harvey Lectures*, 77:43–62, 1983.

In Quest of Fleck

Fleck, L. *Genesis and Development of a Scientific Fact: Introduction to the Study of Thoughtstyle and Thoughtcollective*, eds. T. J. Trenn and R. Merton. Chicago: University of Chicago Press, 1979. (first German edition, 1935).

Fleck, L. "Specific antigenic substances in the urine of typhus patients." *Texas Reports on Biology & Medicine*, 9:697–708, 1947.

Fleck, L., and Z. Murczynska "Leukergy," *Texas Reports on Biology & Medicine*, 9:709–734, 1974.

Korchak, H. M., K. Vienne, L. E. Rutherford and G. Weissmann. "Neutrophil stimulation: receptor, membrane, and metabolic events" *Federation Proceedings*, 43:2749–2754, 1984.

AIDS and Heat

Barre-Sinoussi, F., J. C. Chermann, F. Rey, et al. "Isolation of a T-lymphotrophic retrovirus from a patient at risk for acquired immune deficiency syndrome (AIDS)" *Science,* 220:868–871, 1983.

Fauci, A. S., A. M. Macher, D. L. Longo, et al. "Acquired immunodeficiency syndrome: Epidemiologic, clinical and therapeutic considerations." *Annals of Internal Medicine,* 100:92–106, 1984.

Gallo, R., S. Z. Salahuddin, M. Popovic, et al. "Frequent detection and isolation of cytopathic retrovirus (HTLV-III) from patients with AIDS and at risk for AIDS." *Science,* 224:500–502, 1984.

Steere, A. C., R. L. Grodzicky, A. N. Kornblatt, et al. "The spirochetal etiology of Lyme disease." *New England Journal of Medicine,* 308:733–740, 1983.

Zinsser, K. *Rats, Lice and History.* New York: Little, Brown & Co., 1935.

Auden and the Liposome

Auden, W. H. *About the House.* London: Faber & Faber, 1965.

Auden, W. H. *Collected Poems of W. H. Auden.* New York: Random House, 1945.

Auden, W. H. *The Dyer's Hand.* London: Faber & Faber, 1963.

Auden, W. H. in: *The Place of Value in a World of Facts: 14th Nobel Symposium,* A.W.K. Tiselius and S. Nilsson. New York: John Wiley Interscience, 1970.

Bangham, A. D. *Liposome Letters.* London & New York: Academic Press, 1983.

Bangham, A. D., M. M. Standish, and G. Weissmann. "The actions of steroids and streptolysin S on the permeability of phospholipid structures to cations." *Journal of Molecular Biology,* 13:253–259, 1965.

Carpenter, H. *W. H. Auden: a Biography.* Boston and New York: Houghton Mifflin, 1981.

Connot, R. E. *Justice at Nuremberg.* New York: Carroll & Graf, 1983.

Diderot, D. *Rameau's Nephew,* trans. J. Barzun and R. U. Bowen. New York: Doubleday/Anchor, 1956.

Forster, E. M. *Two Cheers for Democracy.* New York: Harcourt Brace & Co., 1951.

Hilberg, R. *Documents of Destruction.* Chicago: Quadrangle Books, 1971.

Mitscherlich, A. and F. Mielke. *Doctors of Infamy.* New York: H. Schuman, 1949.

A Fashion in Metals

Ball, G. V. "Two epidemics of gout." *Bulletin of the History of Medicine,* 45:401–408, 1971.

Bywaters, E. "Gout in the Time and Person of George IV: a Case History." *Annals of the Rheumatic Diseases,* 21:325–338, 1962.

Halla, J. T. and G. V. Ball, "Saturnine gout: a review of 42 patients." *Seminars in Arthritis & Rheumatism,* 11:307–314, 1982.

Weissmann, G. and G. Rita, "Molecular basis of gouty inflammation: interaction of monosodium urate crystals with lysosomes and liposomes." *Nature,* 240:167–172, 1972.

The Chart of the Novel

Bertrand, M., in: *Claude Bernard and the Internal Environment; A Memorial Symposium,* ed. E. D. Robin. New York: Marcel Dekker, 1979.

Forster, E. M. *Aspects of the Novel.* New York: Harcourt Brace & Co., 1927.

Zola, E. *Oeuvres complètes* vol. 10, *Le roman experimental.* Paris: Cercle du Livres prêcieuse, 1960.

The Urchins of Summer

Dunham, P., L. Nelson, L. Vosshall and G. Weissmann. "Effects of enzymatic and non-enzymatic proteins on Arbacia spermatozoa: reactivation of aged sperm and the induction of polyspermy." *Biol. Bulletin (Woods Hole),* 163:420–430, 1982.

Thomas, L., R. T. McCluskey, J. L. Potter, and G. Weissmann. "Comparison of the effects of papain and vitamin A on cartilage. i: The effects in rabbits." *Journal of Experimental Medicine,* 114:705–718, 1960.

Weissmann, G. "Changes in connective tissue and intestine caused by vitamin A in amphibia and their acceleration by hydrocortisone." *Journal of Experimental Medicine,* 114:581–592, 1961.

Poussin and the Bomb

Blunt, A. *Art Bulletin,* 20:96–103, 1938.

Huizinga, J. *The Waning of the Middle Ages.* New York: St. Martin's Press, 1949.

Koestler, A. in: *The Place of Value in a World of Facts: 14th Nobel Symposium,* eds. A.W.K. Tiselius and S. Nilsson. New York: John Wiley Interscience, 1970.

Moore, M. *Collected Poems.* New York: Macmillan, 1952.

Panofsky, E. *Meaning in the Visual Arts.* New York: Doubleday/Anchor, 1955.

One and One Half Cultures

Foucault, M. *Birth of the Clinic: an Archaeology of Medical Perception.* New York: Pantheon, 1973.

Trilling, L. *The Liberal Imagination.* New York: Viking, 1950.

———. *Sincerity and Authenticity.* Cambridge: Harvard University Press, 1971.

Turgenev, I. *Fathers and Sons.* New York and London: Penguin, 1965.

Whitehead, A. N. *Science and the Modern World.* New York: Free Press, 1967.

Whitehead, A. N. *The Aims of Education.* New York: Free Press, 1967.

Golems and Chimeras

Auden, W. H. *The Dyer's Hand.* London: Faber & Faber, 1963.

Ayer, A. J. *Language, Truth and Logic.* London: Victor Gollancz, 1936.

Connolly, C. *The Unquiet Grave.* New York: Viking, 1947.

Goodfield, J. *Playing God.* New York: Random House, 1977.

No Ideas But in Things

Coles, R., ed. *William Carlos Williams; The Doctor Stories.* New York: New Directions, 1984.

Mariani, P. *William Carlos Williams; A New World Naked.* New York: McGraw Hill, 1981.

Ober, W. B. "William Carlos Williams: the influence of medical Practice." *Journal of the Medical Society of New Jersey,* WCW Memorial Issue, 80:34–37, 1983.

Williams, W. C. *The Autobiography of William Carlos Williams.* New York: New Directions, 1951.

Williams, W. C. *Collected Early Poems.* New York: New Directions, 1950.

Williams, W. C. *Collected Later Poems*. New York: New Directions, 1951.

Williams, W. C. *Make Light of It: The Collected Short Stories*. New York: Random House, 1950.

Williams, W. C. *Paterson. Books 1–6*. New York: New Directions, 1963.

Couching the Question: Freud Revisited

Freud, S. *The Standard Edition of the Complete Psychological Works*, 12v. ed. J. Strachey and A. Freud. London: Hogarth Press, 1966.

Hyman, S. E. *The Tangled Bank*. New York: Atheneum, 1962.

Malcolm, J. *In the Freud Archives*. New York: Knopf, 1984.

Masson, J. M. *The Assault on Truth*. New York: Farrar, Straus and Giroux, 1984.

Medawar, P. *Pluto's Republic*. Oxford and New York: Oxford University Press, 1982.

No Frogs in Berlin

Brecht, B. *Selected Poems*. New York: Reynal and Hitchcock, 1947.

Stricker, F. "Üeber die Resultate der Behandlung der Polyarthritis rheumatica mit Salicylsäure." *Berlin. klin. Wschr,* 13:1–4, 1876.

Tyler, M. J., et al. "Inhibition of gastric secretion in the gastric brooding frog *Rheobatrachus silus*." *Science,* 220:609–610, 1983.

Überla, K. K. in: *Möglichkeiten und Grenzen der Technischen Kultur,* ed. D. Rossler and E. Lindenlaub. Stuttgart/New York: F. K. Schattauer Verlag.

Vane, J. R. "Inhibition of prostaglandin synthesis as a mechanism of action for aspirin-like drugs." *Nature.* 231:232–235, 1971.

Weissmann, G. "From Auden to arachidonate." *Cellular Immunology* (Festschrift for Lewis Thomas), 82:117–126, 1983.

Nobel Week
Nobel Foundation: Text of citations for the 1982 awards, 23pp.

Against Aequanimitas
Bensley, E. H. "Gertrude Stein as a medical student." *Pharos*, 47:36–37, 1984.

Carrell, A. *Man the Unknown*. New York: Harper, 1935.

Cushing, H. *The Life of William Osler*, 2v. London and New York: Oxford University Press, 1940.

Forster, E. M. *Abinger Harvest*. New York: Harcourt Brace & Company, 1936.

Osler, W. *Aequanimitas, etc.* Philadelphia: P. Blakiston's Sons, 1905.

Osler, W. *The Old Humanities and the New Science,* intr. H. Cushing. Boston & New York: Houghton Mifflin, 1920.

Williams, W. C. *Collected Early Poems*. New York: New Directions, 1950.